EATING THE OCEAN

La collection Louis J. Robichaud/The Louis J. Robichaud Series

Directeurs de la collection/series editors: Donald J. Savoie and Gabriel Arsenault

Collaboration entre l'Institut Donald J. Savoie et McGill-Queen's University Press, la collection « Louis J. Robichaud » regroupe des publications en français et en anglais, évaluées par des pairs et qui portent sur les politiques publiques canadiennes, le régionalisme et des sujets connexes. La collection est ouverte aux chercheuses et chercheurs de toutes les universités du Canada atlantique et offre une plus grande visibilité aux travaux universitaires qui s'intéressent aux enjeux fondamentaux de politique et comprennent des analyses rigoureuses, pertinentes et non partisanes à forte valeur sociale et contribuant à des débats publics éclairés. La collection rend hommage à l'ancien premier ministre néo-brunswickois Louis J. Robichaud, reconnu comme l'un des grands leaders politiques du XXᵉ siècle au Canada et l'un des fondateurs de l'Université de Moncton.

A collaboration between the Donald J. Savoie Institute and McGill-Queen's University Press, the Louis J. Robichaud Series publishes peer-reviewed scholarship on Canadian public policy, regionalism, and related subjects in both English and French. The series welcomes books by researchers at all Atlantic Canadian universities and offers widespread visibility to Atlantic university research that focuses on the stakes of public policy, informed by rigorous, relevant, and non-partisan analysis that is socially engaged and aims to contribute to transparent and informed public debate. The series honours Louis J. Robichaud, a New Brunswick premier widely recognized as a leading light of twentieth-century Canadian politics and a founder of the University of Moncton.

EATING THE OCEAN

Seafood and Consumer Culture in Canada

Brian Payne

McGill-Queen's University Press

Montreal & Kingston • London • Chicago

© McGill-Queen's University Press 2022

ISBN 978-0-2280-1449-2 (cloth)
ISBN 978-0-2280-1598-7 (paper)
ISBN 978-0-2280-1557-4 (ePDF)
ISBN 978-0-2280-1558-1 (ePUB)

Legal deposit fourth quarter 2022
Bibliothèque nationale du Québec

Printed in Canada on acid-free paper that is 100% ancient forest free (100% post-consumer recycled), processed chlorine free

This book has been published with the help of a grant from the Canadian Federation for the Humanities and Social Sciences, through the Awards to Scholarly Publications Program, using funds provided by the Social Sciences and Humanities Research Council of Canada.

Funded by the Government of Canada   Financé par le gouvernement du Canada   Canada   Canada Council for the Arts   Conseil des arts du Canada

We acknowledge the support of the Canada Council for the Arts.

Nous remercions le Conseil des arts du Canada de son soutien.

Library and Archives Canada Cataloguing in Publication

Title: Eating the ocean : seafood and consumer culture in Canada / Brian Payne.
Names: Payne, Brian J., author.
Description: Series statement: La collection Louis J. Robichaud = The Louis J. Robichaud Series ; 2 | Includes bibliographical references and index.
Identifiers: Canadiana (print) 20220278563 | Canadiana (ebook) 20220278768 | ISBN 9780228014492 (hardcover) | ISBN 9780228015987 (softcover) | ISBN 9780228015574 (PDF) | ISBN 9780228015581 (EPUB)
Subjects: LCSH: Seafood industry—Canada. | LCSH: Fishery policy—Canada. | LCSH: Seafood industry—Government policy—Canada. | LCSH: Fishery products—Canada—Marketing. | LCSH: Seafood—Canada—Marketing. | LCSH: Food consumption—Canada. | LCSH: Food supply—Canada. | LCSH: Food supply—Government policy—Canada. | LCSH: Consumption (Economics)—Canada. | LCSH: Consumption (Economics)—Government policy—Canada.
Classification: LCC HD9464.C32 P39 2022 | DDC 381/.4370971—dc23

This book was typeset by True to Type in 10.5/13 Sabon

*For all those searching for more sustainable food policies
so that we can better share our global food resources
more equitably across class, race, gender, ethnicity,
and nationality.*

Contents

Figures

Acknowledgments

This project began as a year-long research trip to Library and Archives Canada in Ottawa, funded by a Hayes-Fulbright grant and Carleton University. During the 2016–17 academic year, I had the great privilege of being the Fulbright Chair of North American Studies in the political science department at Carleton University. I spent most of my time, however, at 395 Wellington Street, where archivists patiently retrieved box after box of government and industry documents. I am thankful for the funding, my faculty colleagues at Carleton University, the archivists at LAC, and the excellent people at Fulbright Canada for making that year the most productive of my career. I also thank the US Embassy in Canada for their support, and, perhaps most importantly, the very fine people of Ottawa, particularly those in the Rockcliffe Park neighbourhood where my family and I lived, where my two sons went to French-immersion school, played soccer, ice-skated, and enjoyed the playgrounds with their new Canadian friends. We all still look back on our year in Ottawa as one of our most joyful years as a family.

The time in Ottawa produced the research; the writing was another hurdle. Bridgewater State University is a teaching school and finding time to research and write is an omnipresent challenge. If not for my colleagues in the Department of History and Canadian Studies, I would not have completed this project. I especially want to thank my friend Andrew Holman, who listened to numerous renditions of my thesis over coffees, beers, and lunches. He took the time to read and edit sections of the manuscript, although I never burdened him with the whole thing. His exceptional feedback dramatically improved the study.

I also want to thank Will Knight at the Canadian Agriculture and Food Museum in Ottawa and Matt McKenzie at the University of Connecticut, Avery Point. Both Matt and Will read sections of the manuscript and provided feedback and encouragement that helped me get to the finish line. They helped me, each in their own ways, to better contextualize the study within the framework of food history and fisheries history. My friends at the Northeast and Atlantic Region Environmental History forum each read an entire chapter and provided valuable feedback. I especially thank Claire Campbell at Bucknell University, Ed MacDonald at the University of Prince Edward Island, and Mark McLaughlin at the University of Maine, each of whom helped me understand this history as a story of modernization in Canada.

I also need to thank the two peer reviewers. Although I suppose I'll never know who you are, your edits and feedback took a messy draft and turned it into a less messy manuscript. The editorial group at McGill-Queen's University Press showed patience and skill in massaging the manuscript into a book.

Finally, and most importantly, there is my family, who never let me become so consumed with history that I forget the joy of living in the present. My wife, Shandra, and our two boys, Ben and Danny, kept me grounded, lifted my spirits with good dinner conversation and lots of play time, and allowed me to disappear for hours and even days at a time to struggle with this study.

EATING THE OCEAN

INTRODUCTION

Finding Fish Customers:
Low Consumer Demand for Fish and
Seafood Products in the Twentieth Century

Many of us have acquired a sirloin steak appetite, and an increased wage which permits us to satisfy that appetite.
"Salt Fish Conference," *Acvo Fisheries News*, October 1923

In 1932, the Canadian Department of Fisheries hired the self-proclaimed "Fish Evangelist" Evelene Spencer to tackle what the department saw as perhaps the greatest threat to the nation's fisheries. That single threat challenged the stability and future prosperity of fisheries from the Atlantic Coast to the Pacific and all the inland waters. It challenged the ground fisheries, the inshore fisheries, large-scale operators, and small single-vessel owners and threatened the livelihood of fishing labourers, managers, owners, and capital investors. The problem was not an invasive species, nor was it the well-documented conflict between steam trawlers and small boatmen in the Atlantic.[1] Nor was it foreign competitors or a polluted and collapsing resource. It was not something that the industry could handle on its own, nor a problem the newly created fishermen cooperatives could address. It was not even production based.

The perceived problem was the dwindling consumer demand for seafood. With the introduction of modern packaged foods, the decreasing cost of meat in all corners of the nation, and rising interest in more nutritious meals incorporating fruits and vegetables, Canadians were turning less frequently to traditional seafood like salted codfish or pickled herring. The fisheries industry needed to modernize its product. It had to be canned, fresh, or frozen, "wholesome," and, most importantly, ready to cook right out of the package. Seafood needed to satisfy the new modern Canadian housewife determined to provide her family with modern, wholesome, inexpensive, and easily prepared meals. Yet instead of focusing on improving the quality of fish products like canned salmon or frozen fillets, both already with well-deserved reputations for poor quality, the Department of Fisheries, alongside the nation's leading fisheries trust, the Canadian Fisheries Association, chose to focus on increasing consumption. Consumer marketing, they believed, held potential for the recovery, stability, and growth of a failing industry. Moreover, in assuming that modern marketing would increase sales, they called for increased production to meet the increased demand. As a result of this thinking, the government agency and its industry partners turned to Evelene Spencer to spark a seafood consumption revolution by convincing Canadians that the nation's domestic supply of seafood, in all its forms, was a modern, wholesome, inexpensive, and easily prepared food ideally suited to the nation's emerging middle class.

Spencer sought to execute this task by turning to the nation's housewives. Her gospel was simple. Seafood was no longer a plebeian dish but had all the essential characteristics to become a staple of the modern middle-class diet. Spencer chastised women for lacking imagination and skills as cooks to prepare appealing fish meals. Canadian housewives must cast aside their frying pan – an obsolete cooking method, according to Spencer – replacing it with her superior Spencer Method of high-heat baking, to produce a pleasing meal with the nourishment demanded of a newly health-conscious middle class. In many ways, Spencer tapped into existing assumptions about food, which by the 1930s had been revolutionized by new discoveries of vitamins and new idealizations of health and quality of life in the modern age. She also tapped into deeply established gender ideals. By the 1930s, Canadian housewives were expected to be informed home economists who did their duty in accordance with scientifically proven methods of managing a modern home, including, of course, preparing modern meals. Within this context, the nation's fisheries bureaucracy and industry leaders sought to save the nation's fisheries by teaching women to better select, prepare, and serve seafood.

This book's main goal is to align the history of fisheries in Canada with the history of food consumerism. Since the publication of Harold Innis's *The Cod Fisheries* in 1940, much of Canada's (and the world's) fisheries history has focused on understanding the fisheries as a resource economy.[2] The focus shifted from macroeconomic histories of the industry to social history of fishing labourers and fishing communities to environmental histories of the oceans, yet few historians have spent much time discussing the consumer product itself. This book hopes to add to the vast historiography on Canada's fisheries by beginning with the recognition that fish, in the end, is food. It asks what strategies the fisheries industry and the fisheries bureaucracy of the Canadian government used to market fish to domestic consumers. Were those strategies successful? Why or why not? What impact did these marketing efforts have on the consumer, the industry, and the resource itself?

This study describes how the fisheries industry and the fisheries bureaucracy of the Canadian government tapped into contemporary food culture linked to new concepts of food health and nutrition to sell more fish to more Canadians by emphasizing the "wholesomeness" of seafood. The marketing mainly targeted women and tapped into early twentieth-century popular conceptions of motherhood. It also emphasized Canadian nationalism, with claims that increased consumption of fish would build the national economy, improve interregional connections, and aid in national victory during times of war. Yet these marketing efforts failed to increase domestic demand for seafood. They failed because both industry and government incorrectly assessed the problem. Whereas they believed Canadians ate less seafood because of poor consumer knowledge of its benefits and housewives' inability to prepare high-quality seafood meals, the failure to increase demand was really owing to the quality of Canada's seafood. Poor extractive methods damaged fish or brought up low-quality species; poor storage, processing, and transportation failed to deliver high-quality products; poor handling practices at the retail level resulted in inconsistent quality. Pumping more money into consumer education had little impact and distracted the industry and the government from the real problem: the mismanagement of the resource and its extraction. Yet throughout the first half of the twentieth century, the industry and the government remained convinced that their marketing was working, or would eventually work, and a mass demand for seafood would emerge on the domestic market. This hope encouraged continued and increased production even as the industry produced more product than the market could absorb. For the resource, this meant continued exploitation beyond sustainability. Although most historians know that Canada's fisheries were over-exploited, this book adds to the story by showing that over-exploitation not only failed to recognize environmental awareness and contemporary scientific knowledge but also did not make basic economic common sense. Even in a pure capitalist society void of moral economy and ecological awareness, the fisheries should have avoided the high level of extraction based solely on a logical assessment of consumer demand. The collapse of Canada's fisheries was not just a story of ecological short-sightedness: it was also a story of failed economic rationalism.

Fisheries historians have played a critical role in efforts to understand just what happened to the world's marine life. Not that long ago, the oceans potentially provided us with an enormous amount of protein, fat, and calories. Today, such hopes are dashed on the rocky realities of seriously depleted biomasses across the oceans. How did we go from abundance to scarcity in such a short period of human history? Fisheries historians have analyzed and criticized ineffective or unenforced policies and laws, misguided scientists, and greedy corporations and fishermen. We have not, however, taken into central consideration a key factor in fisheries exhaustion: that the final product is food. As such, the consumer culture of food, and government and industry policies to manipulate that consumption, played a role in fisheries extraction and eventual exhaustion. This work follows the shifting focus of the Canadian Department of Fisheries and the Canadian Fisheries Association away from production-side solutions to economic decline to a new focus on generating increased consumption as a means to facilitate industry recovery, stability, and growth. Beginning with food-rationing programs during the First World War, the federal government began to assist the fisheries industry in marketing seafood to a wider, more diversified consumer group. They specifically targeted communities far from the ocean and marketed seafood to an emerging white middle-class consumer base. At first, this marketing emphasized the need to replace meat with fish as a patriotic responsibility, but following the First World War, and to a lesser extent, the Second World War, the government continued assisting with marketing to increase fish consumption. Beginning in the 1920s, the Canadian fisheries bureaucracy paid for marketing that emphasized the nutritional value, ease and diversity of use, and cost savings of consuming more fish. The government also assisted the industry in marketing new fish choices like canned pink salmon, canned lobster, or fish and chips to relieve identified stresses on the key species in the market, such as sockeye salmon or haddock.

This effort was the industry and government's proposed solution to what had become for the Canadian fisheries a persistent problem of imbalance between its growing capacity to increase production and a market that maintained stable, if not reduced, demand for the products. As in agriculture, the result was perpetually glutted markets, which had two profound and long-lasting impacts. First, glutted markets drove down wholesale and retail values in proportion to cost of production, reducing capital return for both business and labour. Second, producers sought to increase overall production to balance out the decreasing returns on per-unit sales. This second impact had negative implications for long-term economic viability by further glutting the market but also had long-term implications from increasing exploitive pressure on the environment from which the resource was extracted.

The fisheries industry, along with the Department of Fisheries, sought to address the imbalance between production and consumption not by reducing production or making that production more efficient but by taking steps to manufacture increased domestic demand. More domestic demand would increase overall sales and restore economic viability to the industry and would also partially liberate Canadian fisheries from a long dependency on foreign trade: Canadian fisheries exported over 80 per cent of their product. During the first decades of the twentieth century, especially after the First World War, that dependency became problematic, as many nations began to erect nationalist protective trade policies. To better balance foreign trade with domestic consumption, fisheries industry leaders and government bureaucrats marketed Canadian fish to Canadians. By increasing seafood consumption across the nation, the marketing essentially argued, Canadians would be healthier and more middle class and would aid in rehabilitating one of the nation's most important and historic industries.

The industry-government marketing partnership relied to some extent on a consolidation of economic power within the fisheries. During the first years of the twentieth century, individual Canadian fishing firms began to coordinate their efforts and build trade associations among themselves and with wholesale and retail distributors at both regional and national levels. Fisheries cooperatives like the British Columbia Packers Association and the Canadian Fisheries Association attempted to pool their market strength and bring some order to an industry still dominated by local, relatively small-scale operations. Similar to other food industry associations like the fruit and vegetable cooperatives in California and Florida, these fisheries trusts sought to standardize product and to market generalized, non-specific commodities. Yet, unlike the celebrated examples in California and Florida, Canadian fisheries associations largely failed to standardize the industry, improve the quality of the product, or develop uniform marketing campaigns. Quality-control measures remained voluntary until after the Second World War, when the federal government stepped in to mandate certain quality standards at the wholesale and retail levels. Until then, when one company sought to decrease cost by packaging and selling poor-quality seafood, the entire industry often felt the ramifications of the negative market impact. The variability in quality and general lack of harmony within the industry across the nation made a general marketing campaign nearly impossible as a means to increase domestic consumption and balance out the supply-and-demand equation.

By the mid-1920s, those who sought to bring this kind of uniformity and order to the industry increasingly turned to the government to provide the impetus for the work. Because the fisheries retained not just important economic value but also important cultural-heritage value to Canada, the government felt bound to lend direct assistance to the recovery, stability, and growth of the industry. Until the end of the 1930s, it increasingly became necessary for the federal government, via the Department of Fisheries, to oversee consumer informational marketing and advertising of Canada's fisheries products. As such, the federal government provided public funds to support private industry. Bureaucrats and politicians within the Department of Fisheries justified this spending and intervention on broad economic and cultural grounds, arguing that the nation's fisheries industry was central to the Canadian economy and that its failure would have profound ripple effects across the nation. Failure would be devastating to regions already facing chronic economic depression, but especially Atlantic Canada. Moreover, the fisheries provided a chief export commodity that Canada could not afford to lose if it wished to remain relevant in international economics. But the rationalization went well beyond economics. Lobbyists for government spending argued that the Canadian fisheries retained important heritage values. A nation largely founded on the fisheries could not afford to lose its fisheries. The fisherman carried iconic value worth more than just dollars and cents. Saving the fisheries became a public necessity of nationalistic importance.

Thanks to this lobbying, by the late 1920s the government solution for the perpetual depression of the fisheries increasingly focused on stimulating consumption. The work of the Department of Fisheries during the 1920s and '30s illustrates public efforts to restore collapsing private-sector industries of significant economic, political, and cultural value, via stimulating increased consumption: a consumer-oriented answer to recession. In an example of progressive optimism, the federal government believed it could direct consumer behaviour to achieve a public end. Rather than addressing production-side problems identified and vocalized by both internal and external witnesses, the government focused on domestic consumption. This focus remained powerful until the early 1950s, when new assessments of consumption levels exposed the clear failure of the government and the industry to stimulate Canadians' appetite for fish, despite over thirty years of trying to do so. By this point, government and the industry were instead making a significant transition towards improving the quality of seafood products, thus making the early 1950s a good stopping point for this study.

This government-funded marketing not only failed to achieve its goal of increasing domestic consumption but also failed to improve consumer knowledge of fish and the environment of seafood, and perhaps even undermined existing understandings. To prevent accusations of bias towards any one sector of the nation's fisheries industry or any one type of fish product, government marketing throughout the 1920s and '30s had to present generic rhetoric and imagery of fish. Throughout the internal documentation of these campaigns, Department of Fisheries bureaucrats continually insisted that no specific fish species, product, or environmental source should be highlighted. Instead, would-be Canadian consumers were presented with vague and generic images of fictitious fish and seafood species. This marketing strategy, focused on national industry recovery, potentially warped consumer knowledge of fish and fisheries resources. This generalized, nationally focused marketing also tended to project the idea of abundance and inexhaustibility of the total fish biomass even as scientists and policy-makers had recognized that some specific species were under threat.

Fisheries historians have long discussed the assumptions of inexhaustibility and the interplay of political, economic, scientific, and cultural understandings of fish. Arthur McEvoy's pioneering work set much of the pattern for future fisheries historians to follow. McEvoy showed how the ecology of fish, market forces, legal rules, and social and cultural forces interacted to shape public perceptions of fish and fishery extraction: "The resources themselves were passive objects of technology and political manipulation, while harvesters' freedom to use fisheries as they saw fit was an article of faith."[3] In this way, fish were malleable and could be used by scientists, industry, and government to achieve specific ends. Such concepts were not in play just for extraction but also influenced consumption. The present work builds on that concept to show that government bureaucrats and industry leaders could also use fish to achieve social objectives at the consumer end of resource economics. Their marketing of seafood between 1900 and 1950 pushed forward the idea of modernity and a more holistic control of nature. This history then adds to others' work on the control of resource extraction within modern industrial societies by also showing how societies (Canada, in this case) also sought to mold a resource (fish) to fit modern consumer ideals (seafood). While the introduction of new catching methods and technologies industrialized the capture of fish, new methods and technologies of transporting, marketing, and selling that fish held the promise of industrializing its consumption. To do so, the selling of fish had to modernize and embrace the revolutions in marketing and advertising that took place early in the century.

This modernization of fish marketing and advertising touched on dominant cultural and political themes of the time and varied little from established marketing ploys within the larger food industry. Such consumer campaigns touched on health, nationalism, war, and economic recovery. At the core of the campaigns, the department promoted fish as a form of health food to counter growing concerns of a malnourishment crisis. During the 1920s and '30s, doctors and nutritionists embraced new discoveries of vitamins and food health and utilized food surveys to expose what they believed was Canada's hidden hunger (micronutrient deficiency) crisis. The Department of Fisheries and the industry argued that fish was a clear solution, and by increasing fish consumption, Canadians could improve their own health as they improved national health and well-being. In this regard, government-funded advertising of Canada's fisheries became a kind of public-health awareness service.

The campaigns also constructed an idealized and simplistic version of gender that buttressed Canada's burgeoning national identity and a modern consumer market. The 1920s and '30s were central decades in the construction of a national identity that was limited to a white idealization of a Canadian middle class. Fisheries advertising celebrated the nation's housewives as the bedrock of an emerging Canadian middle-class ideal. As central actors within the family, women were called upon to provide healthy, varied, and stabilizing meals to fuel social and economic growth. This Canadian middle class was projected as universally white. No Indigenous people, Afro-Canadians, or any other ethnic group – other than Jewish populations targeted in religious-themed advertisements – appear in Canada's seafood marketing before the 1950s.

The Department of Fisheries felt direct government support of a private-sector economy was justified by the persistent economic decline of the industry. For the fisheries, the Great Depression had started in 1919. With the end of wartime rationing, many Canadians quickly switched from fish back to more desirable meat products. Despite maintaining high levels of production, the fisheries industry struggled until 1939. Although some witnesses, including fishermen, a Royal Commission, and a government investigation into price spread, suggested that the key to industry recovery was improving final product quality, the Department of Fisheries embraced industry lobbyists' claim that the real need was to increase consumption. Thus, during the 1920s, and especially during the '30s, the government tried to use increased consumption as a tool to elevate economic depression. The effort's failure, both in theory and in execution, was not exposed until the early 1950s.

Behind the scenes within the Department of Fisheries and between the government and industry leadership, the main justifications for the consumer campaigns were national food security and economic stimulus. During both world wars, consumers were instructed to eat fish as a form of patriotism and direct engagement with the war effort. By switching from meat to fish, they conserved meat products for export to fellow Canadians on the war front and Allied civilian populations bearing the brunt of the conflict. Yet these patriotic messages were balanced by internal communications that focused on ways to use an immediate mandatory shift in consumer habit to promote long-term economic growth well beyond the war years.

By 1952, however, it was clear that the problem was not one of quantity of consumption but of quality of supply. During the first four decades of the twentieth century, those involved in aggressive marketing simply assumed – with no firm data in support – that marketing would inevitably result in increased consumption. The Department of Fisheries and industry leaders thus sought to increase production to meet higher levels of demand. They particularly sought to increase production of the new fresh and frozen fish trade largely supplied by the nation's young Atlantic trawler fleet. Increasing production, however, only further complicated the problem. It increased supply in an already glutted market and did so with what many viewed as an inferior-quality product, warping consumer appreciation for it and further depressing demand. Increased production, of course, also added to pressure on the environment from which the product was drawn. The fishing industry continued to pull more fish from the ocean at a level that neither the ocean environment nor the fish market could sustain. In hindsight, it would appear as if economic and environmental collapse was at least partially due to gross miscalculation of consumer demand and disregard for calls to address production-side problems.

This history begins with the age of consumer capitalism, which began in earnest during the last two decades of the nineteenth century, though the fisheries industry did not delve too deeply into consumer culture until the early years of the 1900s. At that time, seafood producers and sellers developed more sophisticated advertisements for fish that shifted from simple information about supply and price towards crafting consumer demand or knowledge of the product. By about 1910, the fisheries industry began to understand that it competed in a diverse and ever-expanding food market in Canada and needed to recast the image of fish to make it competitive with a wide variety of other foods. Fish could no longer be the food of the poor. Cheap prices would not bring in more customers; the industry had to start catering to the middle class. By the end of the 1910s, and certainly by the onset of the First World War, the Canadian federal government began taking more decisive steps to aid the industry in marketing seafood and developing a more robust and sustainable demand for Canadian fish.

The book carries the history of seafood consumerism through to the end of the Second World War. There is little doubt among historians, scientists, and policy-makers that the beginning of the end for the world's fisheries occurred sometime between the end of the Second World War and the 1980s. Most argue that the peak of global fisheries production occurred in 1996, with an official global catch of 86.4 million metric tons but potentially as high as 130 million metric tons, depending on how the data are worked.[4] But overfishing had begun long before, with many key fish stocks starting to collapse between the 1930s and '60s. Pacific mackerel collapsed in 1933, California sardines in 1949, Norwegian and Icelandic herring in the 1950s, South African pilchard in 1960, Peruvian anchoveta in 1972, and George Banks herring in 1967.[5]

Most of the postwar accelerated growth in global fishing power was held by those nations with well-developed fishing fleets, which increasingly turned their attention to the waters of traditionally non-commercial-fishing coastal nations. These fleets were built, or rebuilt, on government subsidies to encourage increased fish production largely for political reasons associated with postwar and Cold War geopolitical strategies. The ability of these newly enlarged fleets to exploit the waters of underdeveloped nations came as a result of a series of international agreements during the 1950s.

Global fishing efforts fundamentally changed around the 1950s to become more international, meaning that the more-developed nations increasingly exploited fisheries outside their territorial limits. Important questions remain as to how Canada participated in this new reality. Although historians have shown that the United States, Great Britain, Japan, Spain, and the Soviet Union were all eager to expand their international fleets, Canada was perhaps more exploited by this development than able to take advantage of it. The nation's deputy minister of fisheries, Donovan B. Finn, for example, was one of the few fisheries scientists or policy-makers who opposed the theory of Maximum Sustainable Yield and its application to international fisheries management.[6]

Not only did the act of fishing – that is, the production of seafood – change around 1950 but fisheries products, the consumer good itself, fundamentally changed as well. When the frozen fish stick was introduced in 1953, followed by cheap, nondescript frozen fish sandwich patties, consumption of seafood changed forever. Fish sticks were made from large blocks of fish flesh, largely caught in Canada, Norway, and Iceland, then breaded, fried, and refrozen. The US government spent money on consumer educational campaigns to introduce this new commodity, and the program was enormously successful. For entire generations of Canadians and Americans, seafood became synonymous with fish sticks. This is a different history, and one that has already been explored to some extent.[7] Yet Carmel Finley maintained (much as this book argues) that while the US government saw the fish stick as a grand solution to the fisheries question and the collapsing American fishing industry, "the problem was not lack of markets."[8] A market- or consumer-oriented solution to fisheries decline was still the dominant focus of government policy into the 1950s (at least in the United States) but does not need to be deeply explored here, thanks to the work of other scholars. Furthermore, by the early 1950s the Canadian Department of Fisheries had abandoned its emphasis on increasing domestic consumption and turned its attention more to the problem of improving the quality of seafood products. Policy-makers, bureaucrats, scientists, fisheries executives, and fishermen all began refocusing their efforts on production-side problems.

Moreover, most of the post–Second World War fishing effort focused not on human food but on the production of fish meal to feed growing global demand from the poultry and livestock industry. Although there was much rhetoric between the end of the Second World War and the global food crisis of the early 1970s about the potential of the world's fisheries to provide a massive amount of protein, in reality that protein largely went to chickens and cows, as well as many, many domestic pets. Fish as a human food played a decreasing role in why and how people went to sea in fishing boats.

Ending this history in the early 1950s largely misses the place of Newfoundland in the story of seafood consumer culture in Canada. Newfoundland, of course, did not join Confederation until 1949 and before that was not directly affected by Canadian government policies to increase domestic consumption. On the contrary, economic recovery policies oriented around domestic consumption often went hand in hand with tariff protection against imports, which of course affected Newfoundland fish exports to Canada. Geographically, this study also only indirectly touches on similar themes in British and United States history. British policy-makers seem to have begun earlier than policy-makers in Canada to think about fish as a food product rather than just another natural resource. Thomas Huxley's famous 1866 *Report of the Commissioners Appointed to Inquire into the Sea Fisheries of the United Kingdom* made frequent references to fish as a necessary food product for the poor and the need for the government to facilitate increased fisheries production as a means to address domestic food security for Britain's working class.[9] A similar food-policy trajectory occurred in the United States. The American food market was obviously much larger than the Canadian, and contemporary observers argued that the US population had slightly higher per-capita fish consumption than Canada, although such data are hardly definitive by today's standards. Yet, unlike Canada's, the American fish market relied heavily on cheap imports, a large proportion of which came from Canada. So, while Canada was largely an exporter of fish, the United States was an importer. As such, due to different domestic economic and political contexts, the story of seafood consumer culture and policy unfolded differently in Newfoundland, Great Britain, and the United States than it did in Canada.

Why is a history of the culture and politics of seafood in Canada now relevant? Seafood has long played a crucial role in global sustainable and secure sources of food. Throughout world history, fish has continuously contributed a large share of world's supply of calories, fats, and proteins. Yet when compared to the proliferation of food history in recent decades, that of seafood remains surprisingly understudied.[10] Food history in the United States and Canada overwhelmingly focuses on agriculture. Histories of wheat, rice, corn, salt, sugar, beef, pork, dairy products, and fruits and vegetables command significant shelf space. In this process, food historians have told us much about the transformation of environments, labour of extraction, industrialization of production, strategies of marketing, culture of consumption, and science of genetic modification of a host of agriculturally produced commodities.[11] Yet with a few exceptions, seafood has not found a place among these works.

Seafood has been surprisingly neglected by fisheries historians as well. While there are many excellent histories of the world's fisheries, few begin from the perspective that fish is food.[12] Most explore the work at sea, the laws and politics that governed fishing, the environmental damage that overfishing wreaked upon the oceans, or the importance of fish to regional economic and social development. The simple fact that many men went "down to the sea in ships"[13] to acquire food for themselves or others is frequently overlooked in maritime histories.

Atlantic Canadian fisheries historians have demonstrated the centrality of the fishing industry to local economic and political policies and to social and cultural legacies.[14] In New England, historians also discuss the importance of political influence and economic growth in fisheries production, at times moving into more critical assessments of the environmental implications of fisheries.[15] Similar histories have unfolded for the Pacific Ocean, although many of those take a more industrialist and "wild west" assessment of the

boom-and-bust of the Pacific fisheries.[16] The vast majority of historical literature of the fisheries focuses on the eighteenth and nineteenth centuries. Some writers are captivated by the idealization of
the "age of sail" and dory fishermen. Others see the colonial fisheries
as the preface to global capitalism, while still others are drawn to the
importance of uncovering the transition from local, shore-side and
sail-powered fisheries of the earlier period, sometimes idealized as
being sustainable, to the industrialization of modern fishing fleets
in the latter decades of the nineteenth century. Few begin with
the position that fish, in the end, is food and is thus shaped as
much by consumer-side economics, with all the attendant cultural
and political influences, as it is by production-side economic and
ecological factors.

Although Carmel Finley focuses on the second half of the twentieth century, her analysis of policy and fish food production is probably the most important historiographical foundation for this study.[17]
Her two books are about the intersections of science, environment,
and state power, but more than most, she addresses the key issues of
consumerism, if briefly. In both books, Finley shows how government
subsidies, particularly in the United States and with US-backed subsidies in important postwar allies, developed a fishing industry "far in
excess of the ability of fish stock to reproduce."[18] Yet not only was the
fishing industry extracting more fish than the environment could sustain: it was also, as this book shows, extracting more fish than the consumer market was willing to absorb, despite industry and government
efforts to increase demand.

Finley continually refers to "cheap fish," especially as it relates to the
Pacific salmon and tuna industry, which she characterizes as "ubiqui-
tous cheap foods." She further notes that the canning of salmon
turned it from a subsistence food for First Nations and Native Americans into a "global commodity with unlimited demand."[19] In fact,
demand was limited, and markets in the Pacific canned-salmon industry were glutted as early as the 1910s. A closer examination of the
creation (or attempted creation) of consumer demand might suggest
that few people wanted to eat fish, and in fact the consumer demand
for fish was likely significantly lower than production levels.[20] These
glutted markets would have developed even earlier had it not been for
the artificial market demand created by world war. As canning did
not start until the 1880s, one might argue that the industry glutted
the market – if we disregard the influence of surging wartime demand

– in fewer than thirty years. Finley also states that "America appeared to have a bottomless appetite for tuna."[21] While it might have been bottomless for Americans and tuna, the same certainly wasn't true across the North American fisheries.

This book shows that, much like production, consumption of fish relied heavily on government subsidies. This resulted in an artificial, or manipulated, market. But, despite the effort, the government subsidization of market demand failed. Instead of increasing consumer demand, it fostered false hopes that aided industry efforts to increase catches and decrease regulation. From this perspective, it is difficult to ignore consumerism's impact on fisheries exploitation. In ways similar to other food histories, this work defines consumerism not just as the effort to promote the interest of consumers in any one industry but also as the broader social and cultural influences that shaped consumer behaviour, as well as political and economic policies designed to manipulate or change that behaviour. A false sense of consumer demand and government policy to increase it contributed to over-extraction of the resource for the production of a commodity that few people wanted.

Finley's work demonstrates with insight, strong documentation, and engaging narration the implications that government policy, specifically subsidies, had on overfishing and the ecological disaster that overfishing created. Yet the twentieth century was not just problematic on the production side; consumerism was equally dependent on government subsidies and warped as much by unsubstantiated and misguiding calculations. Whereas Finley shows how politics and economics warped scientific understanding of ecosystems and biomass, this book shows how politics and economics warped proper assessments of consumer demand. Just as Maximum Sustainable Yield gave a false sense of ecological security, so government-funded advertising led to a false sense of consumer security. Canadians caught more fish than the environment could sustain; they also caught more fish than the market demanded.

While the critique that historians have failed to see fish as food may seem academic, even fishermen did not see fish as essentially food. Certainly they knew that the end product of their labour would end up on someone's plate, yet for most fishermen, catching fish was market-oriented labour, and fish were mostly commodities. Seafood was early seen as a market commodity in a way that other foods were not so rapidly commodified. While we often speak of family or self-sustaining farms, we never talk of such subsistence fisheries beyond the scholarship of Indigenous and First Nation communities.[22] Certainly farmers turned to local waters to augment their food sources, but it would be inaccurate to suggest that those who were primarily fishermen conducted labour that was primarily household oriented. Instead, almost all fishermen across time and space have been more closely tied to a market for fish in ways that many farmers were not. As such, almost from the beginning of humanity's efforts to catch fish, there has been an effort to preserve that catch for market. While seafood can be an excellent source of calories, fats, and protein, it spoils quickly. Processes like smoking, drying, or salting were almost immediately applied to the catch in a way that might be unique for speed of application when compared to other forms of food production.

The ties between politics and culture are essential in this study. It was the politics of industry-government partnership that launched the mass-marketing campaigns of the early twentieth century, and while political lobbying did not result in increased fish or seafood consumption in Canada, it may have altered the economics of its production and the culture of its consumption. Economically, the level of mass advertising that took place after about 1920 required a scale of corporate integration and industry-government partnership not often associated with the fisheries industry. Other historians have begun to show that in the United States the fisheries industry became organized into large, modern corporate structures similar to other industrial growth occurring in the late nineteenth and early twentieth centuries. The new "cooperatives," a 1920s label for a trust, tried to consolidate purchasing, distribution, marketing, and retailing power. Canada, however, generally lacked such consolidated movements, except for some regional groups in British Columbia and the national organization the Canadian Fisheries Association (CFA), which had only limited control over production and distribution. The CFA made headway in organizing the industry only after government partnership during the 1930s and '40s. Among other things, it used this consolidation within the industry and between industry and government to break down consumer knowledge and streamline purchasing of generic products not directly associated with the environment from which they came. The government marketing of the 1930s explicitly downplayed information about specific fish species, industry sectors, and environmental sources in an effort to market a generalized fish product. As a result, this history of seafood consumption makes few references to specific species or environmental information.

Culturally, then, the act of eating seafood fundamentally changed. Prior to mass marketing and distribution of indiscriminate seafood, eating seafood required some knowledge of the type of fish and geographic proximity to the source of production. Buyers bought a fish that looked and smelled like a fish – something of major concern for bureaucrats and industry leaders during the 1920s and '30s who feared that the "ugliness" and "smelliness" of fish limited its consumer appeal. They thus sought to create products that looked and smelled less like fish. Meanwhile, advances in freezing and canning technologies allowed for fish consumption not only year round but also by people who lived far from the environment from which the fish came.

By recognizing that fish is a food, fisheries historians can better understand the causes of resource depletion. Historians' failure to recognize the role that efforts to increase consumption played in fisheries depletion has resulted in limited understanding of the causes of this profound environmental disaster. By seeking to broaden and diversify seafood consumption after about 1920, government and industry set the stage for a level of extraction far beyond what the environment could sustain and what the consumer market would have naturally been willing to absorb. Though the level of seafood consumption in Canada never came close to the target goals of the marketing campaigns, both industry and government nonetheless sought increased production to meet illusionary goals that only dumped more product onto a glutted marketplace, further depressing that product's value. Ultimately the consumer-oriented focus on industry recovery had long-term negative implications for fishing communities, as the losses from declining profits were pushed backwards onto fishermen.

This study can shed light on food commodification well beyond fisheries. Although fish have unique characteristics that set them apart from other food products, the most obvious being that they are the only wild animals we eat on a large scale, a history of seafood gains much from and adds to the broader understanding of food culture. First, through this study we can explore the impact that ideas regarding food nutrition had on food consumption. Initially these ideas were controlled by biochemists, food scientists, or medical doctors, but industry marketing came to shape (or warp) many public perceptions about food health. Second, the war years brought concerns for food security, and especially for the fisheries, with a heightened sense of nationalism that celebrated domestic resources over imports. Third, much of the marketing of seafood targeted women as the key providers of nutritious food that also buttressed the domestic economy. Government and industry deployed overtly gendered ideas about food security not just for the individual or the family but for the nation and projected those responsibilities onto Canada's housewives. Fourth and finally, this history explores the ways in which government agencies, particularly after 1930, saw increased consumerism as the key element for recovery, stability, and growth of the fishing industry. This consumer-focused economic stimulus policy overshadowed production-side problems, which in the long term had profound negative implications. The debate between consumer-oriented and production-oriented economic stimulus also divided the industry by class, as industry leaders generally favoured consumer-oriented solutions to economic decline, while fishing labourers called for more attention to production-side problems of inequity. An economic, political, and cultural history of seafood thus tells us much about the relationships between people, government and industry leaders, and the food that sustains their existence.

The first chapter of this book seeks to establish the economic and political context of Canada's fishing industry and regionalism at the turn of the century. It introduces the reader to the modern state developed during the First World War and the 1920s and to the struggles of Canada's fisheries, especially on the Atlantic coast, to modernize production and distribution. More specifically, it deals with the perpetual debate around underdevelopment in Atlantic Canada and narrows further to discuss dwindling markets for the region's main seafood product, salted fish. Chapter 2 deals with the dominant theme in seafood marketing during the first half of the twentieth century: seafood as a health food. Marketing fish as a foundation for a healthy diet was, however, more than just about personal health. Marketing campaigns incorporated new concerns of a potential nutritional crisis of malnutrition or hidden hunger in Canada. As such, eating more fish was essential for both individual and national health.

Just as the second chapter illustrates how government and industry tried to recast the product as a modern food commodity, the third chapter shows how the government and industry tried to recast the seafood consumer as a modern customer. Fish advertising matched with new ideals about women's roles in scientifically managing their households. The advertisements cast the Canadian middle-class housewife as a modern, informed consumer looking to provide wholesome, inexpensive, easy-to-prepare meals for her family.

The next two chapters deal with government efforts to stimulate consumption during times of crises. Chapter 4 examines the industry-government partnership during the interwar period and the Great Depression and illustrates how industry leaders convinced government bureaucrats and politicians that consumer-oriented economic solutions to industry depression were better than solutions based on evaluations of production-side inefficiencies. As a result, they placed more emphasis on convincing people to buy more fish than on balancing production to meet existing consumer demand, improving the quality of the product, or improving sustainable fishing practices. The long-term ramifications of this approach became apparent only by the late 1950s, by which time the industry and the resource were well on their way to collapse. Chapter 5 addresses marketing of domestic fish consumption as a patriotic alternative to eating meat during the Second World War. It explores both the public discourse controlled by bureaucracies like the Food Requirements Committee and the internal communications between industry leaders and the Canadian Department of Fisheries to explore how the industry and the federal department used the wartime emergency to foster increased public demand and project long-term industry growth. This chapter also explores the carry-over of wartime rhetoric about food security into the immediate postwar period. Within the new geopolitical context of the emerging Cold War, Canada (as well as the United States and Great Britain) sought to develop a world food order. The Canadian Department of Fisheries and the Canadian Fisheries Association sought (unsuccessfully) to carve out a global place for Canadian fish. In this effort, they carried over many of the themes they used (unsuccessfully) to market seafood in the domestic market into the global marketplace, which was controlled by international organizations like the United Nations Relief and Rehabilitation Association and the United Nations Food and Agriculture Organization.

This history ends around 1952, by which time the Department of Fisheries apparently awoke to the idea that quality of product could contribute more to economic stability than quantity of consumption. This realization fit within the larger context of planned food management that emerged out of the Second World War. Those involved in food rationing and food relief in Canada, the United States, and Great Britain began to articulate that rationing and control of food production, distribution, and consumption had resulted in healthier diets for the individual and resulting improvements in public health and also in a more equitable distribution of food resources within nations and across the world. Thus, expanded food management appeared to solve pre-war concerns around social security and equity, wartime fears of global competitiveness, and anxiety that potential postwar scarcity could result in social and political upheaval. Fish fit within this new paradigm as a food product quickly transferable into needed nutrition and as an underexploited resource across the world's oceans. Within this context, Canadian fisheries management increased its commitment to international cooperation in conservation, equitable sharing of knowledge and resources, and control of quality from point of extraction through production to consumption.[23] The Department of Fisheries largely abandoned the campaign for increased domestic consumption and focused more on quality control via investigation and inspection at all points of production, processing, and distribution.

While consumerism continued to play a role in fisheries management well after 1952, this history shows that, during the first half of the twentieth century, consumer demand did not define production. Instead, producers continuously sought to manipulate demand, as was (and is) the case in nearly every consumer industry. Yet, unlike other commodities sectors, the Canadian fisheries industry with its government allies seemed lost in how to deal with consumerism. They failed to develop accurate knowledge of consumption levels or potential consumer demand and assumed that increased marketing would automatically result in increased consumption, which would stimulate sales and resurrect a dying industry. The resulting failure to understand the proper balance between production and consumption meant that Canada's seafood industry may have been the country's most mismanaged industry of that period. Industry leaders and government bureaucrats persistently miscalculated consumption levels, and their assumptions further depressed both the industry and the environment. Because the fisheries are natural resource-based industries, over-extraction had long-term environmental, economic, and social implications that the industry, government, and communities are still struggling to deal with.

1

The Modern State: Government Agency in Marketing Seafood from 1900 through the First World War

Ottawa's assistance in building a strong domestic market for Canadian fish between about 1900 and 1950 was in many ways an extension of the nineteenth-century National Policy, albeit a shift from production-side oriented subsidies (tariffs for domestic industry) to consumption-side stimulus (advertisements to increase consumer demand for national products). The National Policy that began with John A. Macdonald in 1876 had long been seen as a negative for fish producers because the Canadian fisheries industry relied heavily on exports, particularly to the United States. By building a stronger domestic market in Canada to replace international markets, Canadian fishery bureaucrats and industry leaders hoped to finally fulfill the promises embedded in the National Policy of economic nationalism for fishing communities in Canada, especially those in Atlantic Canada.

From the beginning, however, the National Policy was more about politics than economic reality. Before the 1870s, Canadian tariff policy was mainly about paying off debt. The $134 million debt that the Alexander Mackenzie government inherited from Macdonald forced Mackenzie's minister of finance, Richard Wright, a free trader by principle, to raise tariffs from 15 per cent to 17.5 per cent. Yet Mackenzie and Wright were ideologically opposed to tariffs and would not authorize further increases. In 1876, with the Liberal Party firmly opposed to increased tariffs, Macdonald found the issue he needed as the leader of the opposition. This was good politics, but there was some economic rationale for it too. Macdonald's embracing of tariffs was not primarily the result of a commitment to paying down the debt or balancing the budget but was instead dedicated to protecting domestic manufacturing. The goal was a national economy that was independent of the United States.[1]

What became the National Policy began as a political manoeuvre by Macdonald to cultivate a national sentiment and undermine the Liberal Party, but it quickly emerged as a form of economic nationalism. Although historians have since used the label National Policy to identify federal policies related to tariffs, land settlement, and railroad development, Macdonald only talked about the National Policy as it related to tariffs, and those that targeted the United States specifically.[2] This form of economic nationalism was meant to strengthen ties with the British Empire by warding off continentalism, at the same time enhancing the sense of Canadian nationalism by illustrating the value of the federal government in Ottawa. Nor was Macdonald's National Policy intended to be a short-term solution to the immediate budgetary crisis. By 1879, Macdonald had committed himself to permanent indirect subsidies of industry at the expense of freer trade across the North American border. The National Policy's tariffs may thus have aided manufacturers in central Canada, but they hurt staple exporters such as fishermen in Atlantic Canada by threatening trade with the region's and industry's primary customer.[3] Almost immediately, Atlantic Canadian politicians voiced their objections. In 1886, William Stevens Fielding, the premier of Nova Scotia, argued that Nova Scotia, New Brunswick, and Prince Edward Island should withdraw from Confederation and replace Canadian economic nationalism with American free trade.[4]

Throughout the late nineteenth century, Canadian politicians sought better trade relations with the United States in an effort to assist Atlantic Canadian fisheries, but with limited success, due to the growing popularity of economic nationalism in the United States. In 1888, Republicans in the US Senate rejected the proposed Chamberlain-Bayard Treaty presented by the Democratic administration of Grover Cleveland, due partially to partisanship and partially to the party's commitment to tariffs and protectionism.[5] Atlantic Canada's goal to develop better commercial relations with the United States had to contend with many shifting influences – manufacturers in Central Canada eager to receive trade protection, protectionists in the United States seeking their own economic nationalism, and British foreign policy actors interested in appeasing US interest and reducing their own commitments in North America. Meanwhile, Canadian fishermen eager for better trade terms with the United States were over-

shadowed by other US-Canada-British issues such as the Bering Sea
pelagic seal debate or the Alaskan boundary dispute. This was the
context in which those involved in fisheries management increasingly turned to the idea of facilitating a better domestic market. Relying
on international markets for the nation's fisheries at this time was
especially problematic even before the era of economic nationalism
following the First World War.[6]

Fisheries policy in Canada had always been closely tied to trade
policies with the United States. Support of fisheries often required
support of free trade. By the election of 1896, the Liberals under Wilfrid Laurier embraced economic continentalism by calling for the
reduction of tariffs and the adoption of reciprocity with the United
States. However, with Republicans in power in the United States, such
free trade was unlikely, as the 1897 Dingley tariff clearly demonstrated.
In March 1899, Laurier seemed content with limited trade with the
United States, stating, "We are not dependent upon the American
market as we were at one time."[7] A subsequent commission set up by
the United States, Canada, and Great Britain made headway on a variety of issues, including fisheries and trade policies, but failed to come
to any settlements.[8] Then, in 1906, the new governor general of Canada,
Albert, fourth Earl of Grey, and the new US secretary of state, Elihu
Root, along with the British ambassador to the United States, James
Bryce, embarked on an ambitious course that historians have dubbed
the "cleaning the slate" phase in North Atlantic diplomacy. The intent
was to settle all outstanding issues between Great Britain and the
United States, many of which directly impacted Canada.[9] Fisheries
policy was certainly part of this theme.

When Laurier again took up the tariff issue in 1910, the context
proved hardly more promising than it had in 1897. The Payne-
Aldrich tariff of 1909 had clearly signalled US opposition to freer
trade. Yet various actors in both countries, from American
newsprint customers to Canadian wheat farmers, pushed aggressively for reciprocity. The January 1911 reciprocity treaty mainly
affected natural products, including fish. Despite Canada's long-
sought objective of lowering tariffs, the January 1911 treaty was
dead by September. The causes of the fatality were diverse and
included some economic rationale, such as Central Canadian manufacturers' fears that this was just the beginning of increased free
trade. For the most part, however, the 1911 reciprocity treaty died
due to politics. Many feared it was a first step towards annexation,

or, at the least, a dangerous shift away from Canada's imperial connections towards continental ones.[10]

Through the late nineteenth and early twentieth centuries, Canadian Conservatives, too, had supported reciprocal trade talks with the
United States. Even Macdonald sought out such talks, but in 1911
Robert Borden warned that reciprocity with the United States would
shift Canada away from British imperialism towards American continentalism.[11] Laurier may have recognized that Canadian freer trade
with its southern neighbour made good economic sense, but he
failed to realize the strength of Canadian nationalism among the
electorate – particularly in Ontario, defined as it was by British imperialism. The failure of reciprocity in 1911 meant further struggles for
Atlantic Canada and increasing complaint that Ottawa did little to
help the nation's fisheries. It was within this context of continued
failure to secure long-term stability in international markets for the
nation's fisheries that Ottawa gradually accepted the need to use its
power to expand the domestic marketplace. The war provided an
immediate reason for such government action, but long-term industry stability and the idea of building a national economy had more
lasting impact.

Government efforts to use the power of the state to spur domestic consumption had two dominant rationales. First, during the war,
the federal government viewed increased seafood consumption as
part of the larger food rationing program that conserved meat
resources for overseas troop deployment and European reconstruction. Second, during the interwar period, politicians and bureaucrats within the Department of Marine and Fisheries concluded,
after a good deal of persuasion from the industry, that increased
consumption would stimulate economic recovery of a historically
depressed industry long abandoned by the nation's economic policy, and thus bring about both profits for companies and increased
wages and job security for the labourers. Both rationales represented concerted efforts by the government to use policy as means to
increase Canadian consumption of seafood, thereby raising questions regarding the role of the modern state to control consumer
behaviour and manage economic development.

The role of food rationing by the state during the world wars has
received increasing attention by historians examining the history of
war and society.[12] The standard imagery in the history of food
rationing has mainly been the hardworking (male) farmer or the

patriotic and informed (female) consumer doing what was necessary to win the war. This history projected a false image of individual commitment and success rather than the highly organized industry-backed reality of food rationing. An examination of the documentation shows both the close cooperation between governments' food boards and food industry leaders and the overt expressions of long-term industry profiting opportunities that could result from expanded government control over food habits. Food rationing was primarily the work of big business and, not surprisingly, primarily benefited big business.

The steps taken during the First World War by the Canada Food Board to increase domestic consumption of seafood is a subject that has received little attention from academic or public historians. Yet it is important, because it highlights the close cooperation between government agencies and Canada's fisheries industry, and also the clearly stated profiteering objectives of the Canada's fisheries industry. Although publicly the federal government marketed fish as a replacement of meat on purely patriotic grounds, internal documents illustrate that the real goal was industry growth and long-term profit. Canada's fisheries industry saw the war as an opportunity to expand domestic consumption of seafood and to sustain that expansion well after the war.

To understand this, we must step back to the pre-war years, to understand how the Department of Marine and Fisheries policies fit within the larger context of a reimagined national policy committed to economic nationalism and domestic market growth. In 1913, the department began shifting its perspective from a focus on science and economics of production to include more investigations into the culture and economics of consumption. In the former role, the department had spent much of its time and money in understanding fisheries resources – the fish and the environment – through economic and scientific perspectives. After 1913, however, it began researching the consumer market and developing public demonstrations and exhibitions to inform Canadian consumers of fish's value as food.[13] Thus, even before the war, the primary regulatory agency for the nation's fisheries had begun to shift towards the need to increase consumption of seafood to aid industry growth.

When war broke out, the Department of Fisheries transferred this consumer-focused goal to the newly formed Canada Food Board. The board had a Fish Section, which focused on increasing domestic con-

sumption of seafood. Working alongside it, in alliance with bureaucrats in the Department of Marine and Fisheries, was the Canada Fisheries Association, the nation's largest fisheries industry cooperative. It
claimed that its members controlled an estimated 90 per cent of
Canada's fisheries productive power. The CFA became deeply involved
in the Canada Food Board's food rationing propaganda during
the war.

In general, the history of Canada's First World War food program
focuses largely on the patriotic messaging of the Canada Food Board.
A main means of communicating its rationing ideals to the public
was the Canada Food Bulletin, which made patriotism the key factor
in food rationing. This patriotic rhetoric was likewise used to stimulate increased seafood consumption as a wartime necessity. The Food
Board called upon Canadians to "Back Up the Troops by Substituting" fish for meat, or to "Subdue the Submarine" by eating more
fish.[14] The board's direct-consumer marketing, however, went
beyond patriotic rhetoric and touched on several pre-war messages.
"Codfish Talks" and a "Daily Fish Bulletin" (figures 1.1–1.7), issued as
inserts in newspapers across the nation, included educational material on Canada's fisheries, histories of fish consumerism in Europe,
and information on health benefits. One such advertisement of 21
October 1919 highlighted new freezing technologies the industry
was using to provide a reliable supply of fresh fish; another of 22
October declared that the national fish day was "distinctively Canadian." A bulletin of 25 October noted that fish was consumed by the
"hardiest races."[15]

Patriotism did appear in many of these messages. A "Codfish
Talk" of 22 October 1917 informed consumers that by switching to
fish they would in one day collectively save enough meat to serve
six million soldiers a half pound each. Another ad released on 28
October stated that by eating fish "we are honouring our men overseas." Many other advertisements pulled at patriotism in another
direction, highlighting the important war effort of the nation's
fishermen, who should be rewarded by purchases of their products.
A 31 January 1919 Canada Food Board ad designed to coincide
with the National Fish Day showed three fishermen-filled dories
desperately rowing away from their sinking schooner with an omi-
nous submarine in the background. The text called upon consumers to "Eat fish to-morrow as tribute to our fishermen and a
recognition to their industry."[16] "Remember the Fishermen," a 23

DAILY FISH BULLETIN

PRESERVATION BY
COLD STORAGE.

ISSUED BY CANADA FOOD BOARD.

That cold prevents putrefaction and keeps
fresh fish and other perishable products
from decay is a fact recognized by everybody.
Efficient cold storage permits articles of food
to be kept for long periods and, within certain
limits, there is no essential change or de-
terioration.

The most extraordinary instance of preservation by cold storage was that discovered
in Eastern Siberia, fifteen or sixteen years
ago, by Dr. O. F. Herz of Petrograd—a huge
frozen mammoth elephant. It became embedded in ice, where it remained for over
2,000 years. Its skin, covered with thick
brown hair, had been marvellously preserved.

Accurate investigations by experts prove
that there is no important difference between
frozen fish and fresh fish of the same species
nor is there any loss of nitrogenous elements,
which give fish their chief value as food.
The consumer, after receiving frozen fish
from the retail fish dealer, should defrost
it in cold water and cook it as soon as possible.

Fig. 1.1   "Daily Fish Bulletin: Preservation by Cold Storage," *Canada Food Bulletin*, 12 October 1919. Although patriotism was the main selling point for the
Canada Food Board's rationing program, the fisheries industry was eager to capitalize on the opportunity to develop a market that would continue after the
war. To do so, they sought a broader educational campaign including information on new technologies like cold storage that had improved seafood quality.

CODFISH TALKS.

TODAY'S
MENU
COD & HERRING

TO YE OLDE
ENGLISH
INN

MOYER

ISSUED BY CANADA FOOD BOARD.

One is accustomed to speak of the "roast beef of old England" and we picture our ancestors growing husky and strong on a generous meat diet, but an examination of the account books of noble houses proves that in early times dry codfish and salt herrings appeared much oftener on the bill of fare than did the juicy roast.

Fig. 1.2 Codfish Talks: To Ye Olde English Inn," *Canada Food Bulletin*, 11 December 1919. The Anglo-nationalism of the First World War played into the nation's home-front strategy. Some Codfish Talks reminded Canadians of their English roots and emphasized that fish, not beef, was the main fare of old English pubs.

DAILY FISH BULLETIN

"CALLER HERRING."

ISSUED BY CANADA FOOD BOARD.

"O, buy ma caller herrin'
They're whalesome fare an' honest farin

So ran the old Scotch song about the tasty herring, a fish that does not find as large a market in Canada as it might.

Pickled in salt and brine, herring are put up in barrels, half-barrels and small kegs. Split and smoked they are marketed as bloaters and kippers. Then again, boneless smoked herring are sold in small boxes under the caption of Digby Chickens." Fresh water lake herring are sold as "Ciscoes" after being pickled and smoked. Kippered herring are put up in cans.

Fig. 1.3 "Daily Fish Bulletin: Caller Herring," *Canada Food Bulletin*, 6 November 1919. Daily Fish Bulletins used fish such as herring to illustrate the shared nationality between the United Kingdom and Canada.

DAILY FISH BULLETIN

NATIONAL FISH DAY.

ISSUED BY CANADA FOOD BOARD

Thursday, October 31st, is Canada's National Fish Day. The idea of a national day on which fish should be featured at one or more meals is distinctively Canadian and was inaugurated by the Canadian Fisheries Association in 1915.

If every Canadian eats fish on October 31st, the saving in meats on that day will amount to three million pounds in round numbers—sufficient to provide half a pound of meat per man to 6,000,000 of the Allied soldiers on the Western front. No meat of any kind should be eaten on National Fish Day.

Fig. 1.4 "Daily Fish Bulletin: National Fish Day," *Canada Food Bulletin*, 22 October 1919. Although this advertisement claims National Fish Day as "distinctively Canadian," other nations too focused on seafood in a celebratory way as a means to ration food by directing consumers to substitute products.

DAILY FISH BULLETIN

THE DOCTOR SAYS:

ISSUED BY CANADA FOOD BOARD.

"It cannot be too strongly insisted upon," says Sir James Crighton-Browne, M.D., D.Sc., of London, England, medical adviser to the late King Edward, "that for working people of all classes, fish is an economical source of the energy necessary to enable them to carry on their work; and that it furnishes children and young persons with the very stuff that is needed to enable them to grow healthy and strong."

Fig. 1.5 "Daily Fish Bulletin: The Doctor Says," *Canada Food Bulletin*, 20 November 1919. Health emerged as a key selling feature for seafood by the war's end and would come to dominate advertising campaigns by the 1930s. This Daily Fish Bulletin reflects the strategy of referring to medical experts, by name or more generally, to buttress the idea that seafood was a healthy choice.

HARDY RACES EAT FISH.

ISSUED BY CANADA FOOD BOARD.

Fish is extremely valuable as a food. Contrary to popular supposition, the hardiest races are not necessarily meat-eating peoples. Take the Scandinavians, the Highland Scotch the Japanese and Chinese—all great fish eaters and renowned for their stamina and enduring powers.

According to Sir James Creighton-Browne, M.D., D.Sc., of London, England, medical adviser to the late King Edward: " It cannot be too strongly insisted on, that, for working people of all classes, fish is an economical source of the energy necessary to enable them to carry on their work; and that for children and young persons it furnishes them the very stuff that is needed to enable them to grow healthy and strong."

REMEMBER!
OCT. 31st, CANADA'S NATIONAL FISH DAY.

Fig. 1.6 "Hardy Races Eat Fish," *Canada Food Bulletin*, 25 October 1919. During the war, Canadian Food Board advertising tapped into ethnic stereotypes to sell fish. "Hardy Races Eat Fish" quotes British doctor Sir James Creighton-Browne on fish as ideal food for "working people of all classes," providing "an economical source of the energy necessary to enable them to carry out their work."

CODFISH TALKS.

HUNTING GROUNDS.

ISSUED BY CANADA FOOD BOARD.

The habitat of the Canadian cod is the great shoal waters which lie in the angle formed by the south coast of Newfoundland and the Nova Scotian and New England coasts; also in the enormous area extending from Cape Cod, Massachusetts, to Cape Chidley on Hudson Straits. The cod abounds in depths of from 20 to 70 fathoms and has from time immemorial been caught by the baited hook and line, though some are captured by netting.

Fig. 1.7 "Codfish Talks: Hunting Grounds," *Canada Food Bulletin*, 2 December 1919. Some Codfish Talks adopted an educational tone, supplying geographic and historical information on Canada's fisheries to tie fish consumption to national heritage.

October spot (figure 1.8) urged: support National Fish Day "as a tribute to our fishermen and a recognition of their industry."[17]

Considering the influence the war had on Canadian society, it is directly referenced surprisingly infrequently in Canada Food Board messages regarding fish. Many of them discussed history, such as a bulletin of 28 November 1917 noting that Canada was settled by fishermen. Others focused on health, as in another 28 November message depicting a strong man lifting a one-ton weight in one hand and a bottle of "fish tonic" in the other. A 22 November message illustrated how housewives could diversify family meals with fish. A 9 November message celebrated trawling as the "simplest method of catching fish" and featured a mother and child watching operations from shore. Since most questioned "the indifference of the Canadian public to the palatability and nourishing qualities" of fish,[18] it can be argued that the real objective was to increase fish consumption regardless of the war.

Private-sector companies were quick to latch on to the government marketing. A 1917 advertisement for fish wholesaler John Brown & Company stated, "Families are sincerely doing their bit to conform to the Food Controller's wishes and using fish instead of meat whenever possible. The Boys in the trenches need our Beef and Bacon. Every ounce you don't eat of these needed meats goes to some one fighting for you."[19] Likewise, Connors Brothers of New Brunswick, the nation's only sardine producer, told customers, "To give strict heed in the order of W.J. Hanna, Dominion Food Controller, that Canadians should conserve the meat supplies by making greater use of fish and sea foods, is really a very pleasant duty to those who are familiar with the delicious products put up under the Brunswick Brand."[20] Fish was good for country and good for company.

Shortage of food was less of a problem for Canadians than increasing costs of that food. Inflation had an immediate and visual impact on prices. Bread riots had erupted across the border in Boston, and increased food prices became a lasting theme reflecting the wartime hardships faced by North Americans. By marketing fish as inexpensive, the Canada Food Board and its industry allies could at once provide consumers with information on alternative food choices, illustrate the steps taken by the government to offset increasing food prices, and increase domestic fish consumption. In a 1917 bulletin designed to get Canadians to eat fish to conserve meat for the troops and embattled allies in Europe, the board also mentioned the eco-

CANADA'S NATIONAL FISH DAY TO-MORROW.

Canadian fishing vessels have been sunk by Hun submarines 60 miles from land.   Eat fish to-morrow as a tribute to our fishermen and a recognition of their industry.

Fig. 1.8   "Canada's National Fish Day Tomorrow," *Canada Food Bulletin*, 22 October 1919. During the First World War, Canada Food Board's rationing program linked patriotic duty to eating fish to help conserve meat for the soldiers at the front. The appeals to patriotism also highlighted the courage of the nation's fishermen who put their lives on the line to bring in the catch. In summer and autumn 1918, German U-boats harassed and sank many Canadian fishing schooners.

nomic value of eating fish: "Fish is a cheaper source of protein."[21] Another communication celebrated the fish-and-chip shops springing up all over urban Canada. Although fish-and-chip shops were often associated with low-income consumer habits and low-quality fish, the federal agency instead tried to celebrate them as active agents in the war rationing program. The shops "deserved encouragement of the fish trade and the custom of the public" for providing inexpensive, wholesome food for the working class, "lessening the cost of living without detracting any from the health or strength of the working man and woman."[22]

The marketing of fish to conserve meat and provide value to the home economy continued well into 1918, when many began to see the war as nearly over. In the May 1918 article "Good Deep Sea Fish

Plentiful and Cheap: Consumers Should Demand the Article at Reasonable Price and Conserve Animal Meats for Use of Allies" in the *Daily Colonist* of Victoria, BC, the Canada Food Board informed readers of a new lower price for flatfish, making them a potential alternative for more expensive haddock. "The use of the [flat]fish is urged on patriotic grounds as well as from the standpoint of their being cheap as an article of food," the article stated. "The Food Board has asked for the consumption of fish so as to release all corresponding amount of beef and pork for use overseas. There should be readiness on the part of all to comply with this request."[23]

Fears of food shortages and their social ramifications continued into the postwar period, particularly during the European food crisis and the heightened fear of socialism during the Red Scare of 1918–19. After the Department of Marine and Fisheries assumed its propaganda mission for the Canada Food Board in 1918, one of its first releases warned, "Certain it is that Famine is the foster-parent of Bolshevism. No normal man who has plenty of food both for his family and himself wants to go rioting in the streets. It takes hunger and desperation to create wholesale Bolshevism and anarchy."[24] Fish consumption too came wrapped in the all-encompassing paranoia.

The Department of Marine and Fisheries and the CFA were important partners in the Food Board's efforts to portray fish as an important part of the patriotic consumer's diet. In 1918, the department provided additional information for the Canada Food Board's new fish cookbook. It had released one in 1913, but the Food Board felt it had covered too much technical information on the fisheries and wanted a slimmed-down version that focused more on national interests, health, and simple and practical recipes. In its introduction, the 1918 booklet called upon readers to "remember the needs of our soldiers and allies and at the same time help a Canadian industry," and to "conserve the land products by eating the products of the sea." The department and the CFA again partnered with the Food Board to reintroduce National Fish Day, a pre-war legacy of the department's campaign to increase seafood consumption. Now, within the context of the war, the focus was on the patriotic necessity of eating more fish. In an early press release, the Food Board claimed, "It is our duty to send overseas every ounce of meat that is possible. This condition will continue for years after the war. Fish is the natural substitute for meat and if it were generally used by our people, vastly larger quantities of meat could be released for shipment overseas."[25]

Assuming that the Food Board's massive wartime advertising campaign would increase consumption, the department also sought to utilize patriotism to encourage increased fisheries productivity. In a 6 August 1917 open letter to the industry, Deputy Minister John Douglas Hazen wrote, "To those who are still in the producing field this condition [of potential food shortages] should appeal as a clear and highly patriotic call for increased and sustained efforts on their part to add to the common food supply." In the letter, Hazen also noted the important business opportunity now present: "It will be readily realized from these facts that, apart altogether from the question of patriotism, an unequalled opportunity at present exists for expanding the Canadian trade in fish and securing new business that, with intelligent care, may be retained after the war is over and world conditions are back to normal."[26] Much of this call to increase production lent support to the nation's few steam trawlers operating in the Atlantic. Trawler operators argued that only via this new technology could the industry provide the volume and consistency of fish required for increased consumer demand. The Canada Food Board celebrated the productive capacity of the trawler. A Canada Food Board film featured trawlers working the Grand Banks, "while German submarines were sinking fishing vessels off the Nova Scotia coast," highlighting the "hazards and chances of the fisherman's life at sea."[27] This film was screened throughout the country, offering Canadians patriotic images of the trawlers and the men who worked on them.

In a similar effort to increase production, William Fisher, assistant superintendent of fisheries in the Atlantic region, called upon his inspectors to enlist the support of local clergy to "address the fishermen of their districts urging increased and sustained efforts to secure larger catches of deep sea fish." Most of the letters in reply from local inspectors, and a few from clergymen themselves, dismissed this as ridiculous as fishermen were already working as hard as they could; any suggestion that they were not was insulting given the gravity of the food situation.[28]

The wartime Food Controllers Office created a Fish Committee specifically to oversee seafood and fresh fish consumption as a wartime necessity. This work, however, was not isolated from larger pre-war, and later postwar, efforts by the industry, in cooperation with the Department of Marine and Fisheries, to expand fish consumption for economic necessity. As the war moved towards an end, the interest of the industry began to shift more to the forefront of communications. A 1917 memorandum from G. Frank Beer and R.Y. Eaton of the Fish Committee of the Food Controllers Office to Charles C. Ballantyne, minister of Marine and Fisheries (1917–21), noted "the urgent need for more closely linking together all matters relating to the Conservation, Production, Transportation, Marketing and Consumption of fish and fish products."[29] Beer and Eaton articulated the view that the Fish Committee had improved transportation, stabilized retail prices by stabilizing supply lines, and educated retailers on the importance of proper storage and display. They worried, however, that Canadians had increased their fish consumption only because the food controller had mandated it via meat rationing, and that once such regulations were removed, fish sales would plummet. The only way to maintain higher levels of fish consumption, they argued, was to reduce the retail price.

Ballantyne responded, "There is no question as to the wisdom of closely linking up conservation, production, transportation, marketing and consumption of fish" so as to reduce retail prices; the Department of Marine and Fisheries would continue this effort throughout the war and after, if necessary.[30] Apparently, those in policy-making roles, such as the department and the Food Controllers Office, saw wartime rationing as an opportunity to expand fish consumption by using high demand to improve industry efficiency and capture a permanent market. Although patriotism was clearly an important factor in the wartime food-rationing program, and the increased consumption of seafood was part of that program, in terms of other agricultural food-rationing programs of the war, historians seem to have stopped at the patriotic story and failed to delve more deeply into the industry's backing of food rationing and focus on the long-term profiteering involved in it. This rationale is clearly illustrated in postwar communications between the Canadian Fisheries Association, the Department of Marine and Fisheries, and the Canada Food Board.

The Food Board's fish propaganda campaign was run by Frederick Wallace, a leader within the CFA and editor of its professional journal, *The Canadian Fisherman*. In a postwar memorandum to the

Department of Marine and Fisheries, the primary purpose of which was to encourage the department to pick up the Food Board's campaign, Wallace stated that the "principal line of effort in propaganda work was the task of educating the Canadian public to eat more fish." He went on to outline the print media campaign, the two films that he directed, and the fisheries exhibits at the Canadian National Exhibition in Toronto. He also discussed at length his work with wholesalers and retailers to cultivate consumer desire through improved signage, display, and handling practices. Although "the peak of the effort to build up a home market for fish has been passed," he maintained it was "essential that the work be carried on."[31] He never explained, however, why such efforts needed to be continued now that the wartime need to relieve consumer pressure on meat had passed.

Wallace called for increased effort to market fish to schools and children, to support low-cost fish-and-chips restaurants, and to reclaim the foreign market. Again, despite his well-articulated plan, he did not provided a rationale but simply assumed that the government would continue its propaganda work in support of the industry. His memorandum, originally penned for Henry Broughton Thomson, chairman of the Canada Food Board, was forwarded to George S. Desbarats, the deputy minister of Naval Service (which had subsumed the Department of Marine and Fisheries during war). In a reply to Thomson, Desbarats wrote, "After further careful consideration of the whole matter, it seems desirable that this department should more actively take up the task of assisting in developing the home markets for fish than it has been doing, and particularly since the creation of the Canada Food Board. Its intimate relation of every branch of the industry, places it in a peculiarly good position to do this. I, therefore, think that it is desirable that the projects to such end, that it had under consideration before the war, had better now be developed."[32]

As the war neared an end, with the public eager to end rationing programs, the Canada Food Board, the Department of Marine and Fisheries, and the Canadian Fisheries Association had realized the need to adopt new marketing strategies to keep Canadians consuming the same level of seafood that they had during the war – if not increasing it. It was not really feasible to continue to push the patriotic message. The Food Board now attempted to argue that the experience of rationing had taught superior food habits that should

be continued in the postwar period. In this vein, the board communicated the idea that rations were "balanced meals" that promoted healthy eating. Fish was a necessary part of this new modern
model of eating. In one remarkable claim, the Food Board stated
that a healthier diet could add thirty years to life expectancy and
was "better than insurance." Thanks to wartime improvements to
transportation, which brought Pacific fish as far east as Winnipeg
and Atlantic fish as far west as western Ontario, more Canadians
had benefited from the "brain-building stuffs supplied by fish diet"
that was part of their rationed diets.[33] By switching from meat to
fish during the war, consumers should have learned, according to a
1918 Canada Food Board message, that fish was "a perfect food; it
satisfies the human palate at times even as meat does not."[34] The
Food Board asked Canadians in 1919, "Why continue to undermine
your health with so much meat? Years from now, vigorous and brisk
from the brain-building stuff supplied by a fish diet, you will wonder how you lived so long without it."[35] Industry allies like Connors
Brothers emphasized in their advertisements the value of heightened consumer knowledge: "The war has been the means of bringing home to Canadians the value of our fisheries. The public is
being educated up to fish, and never before has Canadian fish occupied so high a place in popular favor."[36] It was time for the industry
to transform wartime patriotism into sustained postwar profit. As
fears of food shortage and price inflation dwindled, the new
national justification for seafood consumption was public health.
The fish industry, however, still needed the government to help sustain, and even expand, this domestic market and create for the fisheries a national economy independent from the volatile international marketplace.

Between the development of the National Policy in 1876 and the
beginning of the 1921 recession, Canadian politicians from both the
left and the right sought a difficult balance on economic policy as it
related to the nation's place within the British Empire and its increasingly powerful southern neighbour – which yearly seemed to
increase its share of the Canadian export economy. What most Canadian politicians wanted was a way to protect domestic manufacturing, which usually meant tariffs (even Laurier supported a 30 per
cent tariff on consumer goods), without undermining the nation's
dependence on the export of staple goods – usually to the United
States, the same country from which they wanted manufacturing

protectionism. To do so required trade protectionism on the one hand (for central Canada) and free trade on the other (for Atlantic Canada). Despite the National Policy's effort to facilitate an industrial economy for Central Canada, at the war's end Canada remained overwhelmingly a staple-producing nation, reliant on exports, with only minimal domestic industry.[37] Those industries that did emerge, such as paper manufacturing or processed foods, were largely financed by US capital or were branch plants of existing American industry leaders. This dependency on US capital meant that most of Canada's industry did not generate secondary economic growth but instead relied on the export of raw or semi-processed natural resources. Historians have long debated the reason for such limited industrial growth before the close of the Second World War. Most point to the tradition of and commitment to merchant capitalism as opposed to industrial capitalism, the relative ease and speed of profitable returns on investment in resource-extractive industry, the lack of large volumes of capital to invest in more complex industrial productions, and the lack of government programs or policies to encourage industrial development. Whatever the cause, Canada's economy remained largely dependent on the production of natural resources and the export of those products to international markets.[38] The fishing industry survived so long as it had reliable foreign markets, but in the emerging era of economic nationalism postwar, and the general decline in global fish consumption, Canadian fisheries desperately needed a renewed national policy that would facilitate rapid growth in domestic consumption so as to create a national marketplace for the nation's seafood production.

The deep economic recession following the First World War placed added pressure on Ottawa to do something about what many in Atlantic Canada saw as long neglect of the fishing industry. While the National Policy, it was then believed, aided Central Canada's manufacturing and the settlement of Western Canada, Atlantic Canada received little to no support from the federal government. The interwar period, as David Alexander notes, was "rich in official enquiries" into regional inequalities and economic stagnation in the Maritimes.[39] Although at the time many pointed to Confederation and the National Policy as the root cause of the region's troubles, "Maritime consciousness of economic stagnation and relative decline within the Dominion of Canada only assumed the stature of certainty and reality in the 1920s."[40] While there was always much debate about the relative merit of Confederation in the Maritimes, the region's economic decline during the 1880s and '90s was part of broader economic patterns that affected Canada and the United States and thus not specific to the Maritimes or the fisheries industry. Through his analysis of many key sectors of the Maritime economy, Alexander indicates that the region was able to keep up with Central Canada through the 1880s. In that decade, the Maritimes' growth rate was probably higher than in Canada as a whole and only began to lag in the 1890s. Between 1900 and 1910, Canada's growth rate was 7.3 per cent, whereas the Maritimes' growth rate dropped to 4.3 per cent. Industry-specific and regional-specific economic decline did not begin until the twentieth century and was not fully noted until the 1920s – "an especially bad period for the Maritimes," Alexander concludes. The decline not only affected industry and manufacturing but was most pronounced in traditional resource-based industries such as the fisheries. Between 1911 and 1939, Alexander notes, the fisheries contribution to the Maritimes' economy declined by 8.5 per cent.[41]

Although many in the region, and in the industry, pointed to Ottawa as a cause, historians have shown that multiple factors contributed to the economic stagnation of the Maritimes' fisheries economy. For example, the region had long suffered a general population decline, as Maritimers moved west or, more often, south to the United States. Patricia A. Thornton shows that out-migration from the Maritimes preceded economic decline in the region and was thus more likely a cause of economic stagnation than a result of it. Thornton shows that out-migration peaked in the 1870s–'80s period, when the region's economy was still growing, and, in some cases growing even faster than Ontario's. She concludes that "endemic out-migration from the Maritimes, such as that which occurred in the 1880s, largely anticipated the economic downturn in the region's economy," and further illustrates that the out-migration that occurred at the end of the nineteenth century was largely selective, taking a disproportionate number of skilled urban craftsmen and unskilled female labourers. Thus, out-migration depleted the necessary labour for industrial growth, and the economic stagnation of the Maritimes that came in the twentieth century was not the result, or solely the result, of Central Canada, Confederation, and the National Policy. Instead, it came from a general population loss that resulted in a lack of the labour necessary for industrial economic development.[42]

Not only were Confederation and the National Policy not necessarily the causes of economic decline in the Maritimes but, as T. William Acheson has shown, the National Policy provided important economic growth potential. Tariff protection could help Nova Scotian commercial interests by blocking foreign shippers from the Canadian market. Under tariff protection, the region also had every reason to suspect that it would witness the same growth in textile production that New England saw in the early nineteenth century during its period of economic protectionism. With Canada's only known reserves of iron and coal in the Maritimes, there was reason to assume that the region would become the power centre of Canada's industrial growth. The only thing lacking for the Maritimes to benefit fully from a protective industrial plan, Acheson observes, was access to the capital necessary to finance such operations; the region had to rely on the energy, initiative, and risk-taking of a small group of commercial elite families. Many of these individuals successfully financed sugar refineries, textile mills, iron foundries, and railway centres. All looked promising until 1885; this entrepreneurial spirit, however, faced the daunting context of the global economic downturn of the 1880s that discouraged aggressive risk-taking. In particular, the collapse of the British lumber market undermined one of the region's main economic bases. Exports to Great Britain and the West Indies, in which lumber played a crucial role, helped the region offset its trade imbalance with Central Canada, which, in 1885 amounted to 70 per cent of the trade going from Central Canada to the Maritimes. Furthermore, the shift away from seaborne commerce via the St Lawrence towards overland rail shipping let Montreal businesses ship to the Atlantic via Portland or Boston rather than through the Maritimes. As elsewhere in North America, the recovery from the 1880s' recession resulted in economic concentration as larger businesses were better able to weather the troubled times and buy up assets at low prices and thus extend their economic power.

Acheson notes that, by the 1890s, Maritime entrepreneurs abandoned the new industries (refineries and textiles) and refocused on economies in which they believed they had a natural or geographic advantage, particularly iron and steel manufacturing. Yet by 1900, iron and steel production increasingly came under the financial authority of Montreal and Toronto. Although the Halifax financial sector may have been able to provide the leadership in financial capitalism necessary for regional success, by 1910, or 1920 at the latest, financial leadership was based outside of the region. The result of these shifting economic realities led many Maritimers to increasingly question the value of the National Policy, the financial power of Central Canada, and the whole basis of financial capitalism.[43]

The Maritime Rights movement of the 1920s built upon the nostalgic belief that the region was better off without its Central Canadian partners. Yet instead of rejecting the emerging national marketplace, regional leaders demanded better access to it. Politicians in Atlantic Canada demanded tariff protection for Atlantic Canadian products and subsidized freight rates for shipment to Central and Western Canada.[44] In other words, many people in Atlantic Canada by the 1920s sought increased – not decreased – economic nationalism. By the 1920s, Maritime leadership sought "political solutions to the structural problems created by the National Policy" and "consistently looked to the federal government for aid against all external threats."[45] They wanted a renewed National Policy, and Ottawa was prepared to give it to the fisheries in the form of subsidies for both shipment and advertising. Taken together, what Ottawa sought was an increased domestic marketplace for Atlantic Canada's fish. Yet as Acheson noted of other regional economies, the economic failure of the fisheries was more about structure than policy. Canada simply could not consume that much fish. Nonetheless, the economic leadership of the fisheries, both in the industry and the government, saw increased domestic consumption as the new solution to the perpetual problem of regional underdevelopment.

During the first years of the 1920s, the Department of Marine and Fisheries continued to support the industry's National Fish Day, circulating informational material to newspapers for publication. Jason H. Colon, director of the Publicity, Marketing, and Transportation Division within the federal department, wrote to Frederick Wallace, secretary of the CFA, in January 1921, noting the need to change the focus of National Fish Day to reflect new postwar realities. He wanted to lessen the emphasis on patriotism and increase the focus on fishermen themselves to gain "the attention of our people to the hardships our fishermen face"; he even suggested changing the name to "Fishermen's Day."[46] These departmental campaigns to promote

National Fish Day (which was renamed National Fish Day – Fishermen's Day) focused somewhat on the health benefits of a fish diet, but more, in the words of Minister Ballantyne, on the need to "commemorate the thousands of brave fishermen who daily risk their lives to furnish us with food." In this way, the immediate postwar marketing campaigns remained patriotic, or at least nationalistic, often asking consumers to do their part in the "development of our natural resources." Ballantyne continued, "Each and every Canadian has a share in this great national property and indirectly draws his dividends." [47] In a 1922 propaganda article released to newspapers by the Department of Marine and Fisheries, the government stated that the National Fish Day was to "commemorate these brave citizens of the nation – these indomitable men of the deep, these reserved, staunch and brawny souls who consecrate their lives to the calling." [48] The goal was to convince consumers to use their money to support fellow Canadians, a nationalistic spending policy.

Similar nationalist rhetoric was directed at the consumer in the form of appeals to conserve food and assist national economic growth. Acting deputy director E. Hawken wrote to Dr J. Harold Putman, inspector of schools, noting the impact war had on global food supplies and arguing it was in "the national interest that utilization of fish as a food should be more extensive and more systematic." Hawken asked Putman to increase the use of fish in school programs to aid in these efforts. [49] Deputy director William Found told retailers in a 1921 memorandum that the Canadian people "own a great natural resource" and so "stand responsible for a great economic waste by not making more use of fish in their diet." [50] These internal communications between government bureaucrats and industry representatives often highlighted the idea that it was consumers' responsibility to provide the economic stimulus to rescue an industry that had contributed so mightily to the war effort and national economic growth. Consumers needed to use their purchasing power to promote food security and the economic well-being of the nation. Yet despite this rhetoric of economic nationalism, Canadian fisheries remained very much dependent on foreign markets.

2

Eat Fish for Health:
Nutritional Science
and the Healthfulness of Seafood

Fish has long been marketed as wholesome, healthful, energy-providing food. Today, we regularly hear about the brain-building power of cod liver oil and the essential omega fats found in salmon. Despite this continued marketing, fish remains a relatively small part of the Canadian diet when compared to consumption internationally. Throughout the nineteenth century, with few exceptions, Canadians understood very little about the health qualities of the food they ate. For most, food provided energy and bodybuilding elements that were limited to fats, proteins, and carbohydrates. Late in the century, some began to understand the importance of minerals like salt, but it was not until the early twentieth century that more complex, if still incomplete, understandings of food began to emerge. Biochemists "discovered" vitamins, and medical and health researchers began to develop rudimentary understandings of the links between the vitamins and minerals found in foods and what became known as deficiency diseases, chief among them scurvy, goiter, pellagra, and rickets. Much of this work was done by British and American scientists in underdeveloped regions like Southeast Asia, Africa, and the American South. Although most people had very limited understanding of what vitamins were, where they existed in food, what they did for the human body, and how specifically they prevented diseases, by 1910 there was growing acceptance that vitamins were important. The specific investigation into vitamins and deficiency diseases unfolded within a larger context of public interest in nutritious (more commonly referred to as "wholesome") diets. The food-diet fad emerged in full force by the end of the 1910s and the early 1920s, and has perhaps never really faded.

While biochemists, agricultural scientists, and medical doctors pioneered investigations into vitamins and deficiency diseases, it was marketing firms and food corporations that captured the public's attention on wholesome or nutritious diets. Throughout the 1920s, advertisements largely dominated public perceptions of modern healthy eating. These ads adopted specific scientific "discoveries," or, more often, made general references to "experts," whenever such rhetoric fit the need to sell more product (figure 2.1). The marketing of seafood in Canada was no exception to the trend. Seafood advertisers embraced the nutrition mania and made generalized and often totally unsubstantiated references to the wholesomeness of fish as a food. By the next decade, this health-focused marketing became more sophisticated and began referring to specific elements such as vitamins A and D, iodine, and calcium in the effort to convince people to eat more fish. Furthermore, by the 1920s, the Department of Marine and Fisheries became a major ally of industry in this marketing effort, providing scientific information and stock language to use in marketing, and, between 1936 and 1940, conducting their own consumer-awareness campaign on the health value of fish as food. Despite the general lack of scientific backing for such claims, Canadian consumers increasingly faced an onslaught of marketers pitching the idea that fish was the healthiest base for any diet, providing both children and adults with valuable protein for body strength, vitamins for bone and eye health, and minerals to ward off some of the most feared deficiency diseases of the day.

The historiography of food, along with the subfield of the history of food health, has proliferated in recent years. Even when limited to academic studies, as opposed to food histories produced for a more popular audience, the bibliography of food histories is extensive. Much of the earlier work falls neatly into a commodities history genre, an interesting and useful way to tell a larger story of culture, politics, and economics through the lens of a single food commodity.[1] A large number of food histories take gender as the central interpretive model, while others use food to delve deeply into race and ethnicity.[2] Historians have also explored relationships between food and social security, poverty, and equity, or have used historical debates on nutrition as a means of social criticism.[3] As Canada is a resource-rich nation, and one that especially relies on its ability to produce and export food staples (agriculture and seafood), it is unsurprising that food has played an important role in Canadian historiography and

Fig. 2.1 "Ask Your Doctor!," October 1936, advertisement for Sea Seald Brand, Maritime-National Fish Corporation Ltd, Halifax, NS. To capitalize on growing interest in healthy foods, Sea Seald Brand of Halifax informed consumers on the vitamins in Atlantic seafood. Like other advertising of the interwar period, Seald used the generalized imagery of a medical expert to lend authority to its marketing.

continues to find traction in a renewed interest in culture, politics, and power.[4] This extensive historiography has mapped out for us a solid understanding of the growing realization in the British Atlantic World of the interplay between food nutrition and individual and public health.

Popular discourse on nutrition emerged after about 1920, especially after Benjamin Harrow's 1922 *The Vitamines: Essential Food Factors* and Elmer Verner McCollum's columns in *McCall's* magazine after 1923.[5] In the United States and Great Britain, this popular discourse on vitamins partnered with a larger discourse on inequity and general lack of social security in a modern industrial state. Progressives in the United States launched new food security studies and argued that low-income Americans simply did not make enough to keep themselves properly fed. New understandings of deficiency diseases articulated that although people might not be literally starving to death, poor food health due to poverty created a hidden hunger crisis that had profound social, economic, and political ramifications.

Social service professionals began to link poor nutrition with poor economic output, family instability, alcoholism, community unrest, and even social revolution. In Britain, Sir John Orr, professor of agriculture at Cambridge and later the first director-general of the United Nations Food and Agriculture Organization, and in the United States, Dr Hazel Stiebeling of the US Department of Agriculture, each led large-scale food surveys during the 1930s with clear social-criticism agendas. Orr already had a reputation as a social reformer who linked poor nutrition with economic poverty and inopportunity and demanded social and political reform to address inequality. Orr and Stiebeling and their respective allies saw food health and food security as key elements for addressing social and economic inequality and sought to use their food surveys and subsequent recommendations to usher in a new age of social equity. In 1933, Stiebeling created the first globally recognized dietary standards, based on an "optimum" rather than a minimum dietary intake. Stiebeling sought a diet that would allow individuals to achieve their greatest physical and mental potential, beyond providing the minimum nutrition to keep them alive and relatively healthy. The end result was a dietary standard with especially high levels of vitamin and mineral intake.

In 1936 the League of Nations' Health Organization adopted Stiebeling's standards.[6] These were subsequently criticized as being significantly too high, yet as Ian Mosby has written, the standards set in the 1930s were not the result of ignorance within the profession but instead reflected a social ideal of what "health" was: "At the heart of this was the choice to define [nutrition], not simply as the absence of specific clinical manifestations of disease or illness, but as the achievement of a certain level of 'optimal health' defined much more broadly."[7] For Orr, Stiebeling, and others, optimum health meant a path to optimum social equity. In that view, food could not be understood in isolation from a much larger social problem.

In Canada, this general history of nutrition played out in similar ways. In *Food Will Win the War*, Mosby outlines the steps taken by a growing nutritionist profession in this country to study food health, articulate the causes and ramifications of a hidden hunger crisis, and advise governments on proper policies to address it. Popular, public, and political discourse on food malnutrition blossomed in 1938 with Leonard Marsh's publication of *Health and Unemployment* and the establishment of the Canadian Council on Nutrition (CCN) within the Department of Pensions and National Health. The following year, 1939, the CCN adopted Stiebeling's optimum diet standards for its food surveys. Eventually five such surveys recorded the food health of predominantly low-income families in Edmonton, Quebec City, Halifax, and Toronto (two surveys) between 1939 and 1941.[8]

Under the direction of Earle W. McHenry of the University of Toronto, these surveys concluded that 60 per cent of those studied were malnourished. Although this figure was likely overestimated by today's standards, few nutritionists doubted that widespread malnourishment existed in Canada during the 1930s, and many believed its foundations preceded the Great Depression. Yet as Mosby articulates in his book, the real debate emerged not in the recognition of a hidden hunger crisis in Canada but in the competing explanations as to why it existed and what steps should be taken to address it. McHenry and Lionel B. Pett, chairman of the CCN, interpreted the data as evidence that Canadian housewives did not know how to properly manage their families' diets. It was housewives' lack of knowledge on the finer points of nutrition and their cultural biases regarding food that led so many to feed their families improperly. McHenry and Pett largely endorsed a policy of education to better inform Canadian women how to buy, prepare, and serve nutritious food. Others involved in the surveys, including Dr Frederick Tisdall, a pediatrician at Toronto's Hospital for Sick Children and director of the National Research Laboratories, and Marjorie Bell, a leading food nutritionist and active social reformer, argued that the key cause of malnourishment in Canada was poverty. For these social critics, education would not result in improved nutrition because even if Canadians knew what to buy and how to cook healthy food, they could not afford it – or even obtain it at any cost. Tisdall and Bell endorsed a campaign that would be part of a much broader discourse on social security that did not unfold until after the war.[9]

As important as the academic and political conversation on food health was, it was not food nutritionists and medical health scientists but food companies and their advertising firm partners that dominated public discourse about food health, and their work in this endeavour began long before the economic collapse of the 1930s. Their efforts to "educate" the consumer began in the early 1900s. Although their advertisements often claimed to represent the thinking of experts in the field, the first three decades of food health advertising, from 1900 to 1930, were characterized by general and often unsubstantiated references to a "wholesome" diet. The bulk of this marketing targeted women as the primary audience, with content that largely focused on children's health. As consumer historiography has shown, during this period women took on added importance as consumers and primary agents in shaping the family purchasing decisions; through women, advertisers could target the whole family. Advertisements emphasized the important role women played in shaping their families' health and often laid the whole responsibility for family well-being at the feet of a fairly generic and somewhat simplistic and idealized version of the Canadian housewife.

Like the rest of the food industry, seafood advertisers tried to convince women that their products were wholesome and healthy, ideally suited for family meals, and of reliable, consistent quality. One of the first steps taken was to improve the image of Canadian seafood as a quality food product. At this time, processed and especially canned seafood suffered from a well-established and probably well-deserved reputation of poor quality. The limited consumer demand for less expensive products like canned salmon and sardines often lost out to supposedly superior-quality products from the United States and Norway. The *Canadian Grocer* outlined the implications the Canadian industry's reputation had on sales and profits in a 1904 editorial noting that the British Columbia Packers' Association had closed many of its canneries because there was "no demand for that low quality of fish, and packers lost heavily."[10]

Processed seafood companies like Black Brothers of Halifax, one of Canada's largest and earliest brand-name seafood producers and wholesalers, began to counter this public-relations nightmare with aggressive advertising. Much of it was targeted at the retail grocers and jobbers upon whom they relied for much of their direct-consumer marketing. In this process, seafood-producing companies tried to build brand-name recognition and link their brand to quality. In a 1904 advertisement in the *Canadian Grocer*, the Halifax company stated that its "brands of fish are prepared at our new factory in La Havre, N.S., and are without question the nicest, cleanest and most convenient form in which fish can be sold or used."[11] Convincing consumers that the plant in which the food was processed was modern and sanitary was a key step in convincing them that the final product was of quality standard.[12] Although the federal government licensed fish producers and inspected fish landings, there were few quality control measures or inspection criteria for the final product, as was the case with other foodstuffs in Canada. Thus, consumers were left to evaluate seafood quality based on marketing messages and their own (often unsatisfactory) experiences. When Atlantic Fish Companies of Lunenburg bought out Black Brothers in 1906, they too embraced the new emphasis on brand-name recognition popular throughout the food industry as a mechanism to convince consumers of quality that set the company apart from its competitors. Atlantic Fish's inaugural advertising campaign linked their name to quality: "None but the very best quality of Fish will ever be sold under this label."[13]

Connors Brothers, one of Canada's largest processed seafood producers and the nation's only sardine company, also tried to expand its domestic sales by emphasizing improved quality. In the early twentieth century, the vast majority of Connors' sales, between 80 and 90 per cent in most years, was for export. It embarked on its own domestic advertising campaign in 1906, but it was another five years before the company moved beyond traditional, simplistic advertising messages relaying limited information about product and price. Beginning in 1911, it developed a new campaign declaring that "skill, care and cleanliness are our watchwords throughout the canning process" and offering consumers an "unconditional guarantee of purity."[14] Considering the overall public image of sardines as paupers' food, and the pressure from US and Norwegian imports, this was a bold message to send. In other advertisements, Connors claimed that their "modern, sanitary process of canning" preserved the "natural fish flavour."[15] Ocean Brand Fish also began to emphasize the quality control measures it took to pack their product under "sanitary conditions," resulting in the "rich flavor which has won them such a reputation."[16] These quality messages, however, were limited to the few producers who could establish brand-name recognition. Despite this push to emphasize the quality and safety of seafood, especially canned seafood, consumers remained unconvinced well into the 1930s. For example, a 1932 piece in *Canadian Home Journal* by food columnist Ann Adam (the pen name of home economist Katherine Caldwell Bayley) noted that many readers were "timid about using canned fish freely." Adam was adamant "that canned fish is one of the safest, most wholesome products you can possibly offer to your family!" She emphasized the modern production process and close scrutiny of government and industry inspectors – which, even by 1932, was not exactly true.[17]

Connors Brothers, perhaps more consumer-conscious than any other seafood producer in Canada, aggressively advertised their brand for more than a decade and continually introduced new product lines such as Brunswick Brand, Jutland, and Banquet Brand. The company also developed a new key-opening container and provided grocers with countertop display cartons, strategies in line with generally accepted marketing methods in the food industry but still unusual in seafood marketing.[18] By 1916, "after many years of scientific effort towards the perfecting" of their process, the company claimed success: consumers everywhere were choosing Brunswick Brand sardines.[19] Black Brothers too pushed their brand name in advertising but did not survive the recession years of the 1910s. In British Columbia, salmon packers also developed brand names, but the sheer volume of salmon canners operating there meant their individual efforts were often lost in the sea of over fifty different name brands.

The key to this advertising on quality control measures was an emphasis on the modernness of processing plants. Black Brothers claimed they used "the most modern methods," in processing codfish.[20] Connors Brothers stated, "Our plant is modern and strictly sanitary, one of the largest on the Atlantic Coast," and the "sweet, freshly caught fish are scientifically prepared and packed" in accordance with the company's high standards.[21] Again, brand-name recognition was central to marketing. Customers had to be convinced that one brand of seafood represented a higher standard of quality than some other brand, whether a domestic competitor or a foreign import. This strategy could effectively separate one seafood producer from the industry's general reputation of poor quality. Connors Brothers warned grocers and their customers to be wary of those "whose quality is not proven, whose wholesomeness is not established."[22] Their Brunswick Brand of the "most carefully selected fish, caught in the famous Passamaquoddy Bay" and packed "a few moments after leaving the water" in a "plant at the water's edge" operated "under the most ideal conditions," provided the surest guarantee of a "pure food" product of a quality to satisfy even the most "particular housewife."[23] Throughout their campaign during the 1910s and 1920s, Connors Brother continually emphasized the "very latest and most up-to-date machinery" of their plant and "most expert workmen" who prepared the canned seafood – an interesting claim, considering that by the 1910s most of the sardine workforce was semi-skilled female and child labour.[24] Their "rigid, sanitary regulation," Connors claimed in another ad, ensured that the product was "always pure and wholesome" and "won unlimited praise from Canadian women."[25]

Modern processing was often linked to wholesomeness, which emerged as a key marketing point for seafood. Until the 1930s, most consumers got their information on the health qualities of food from advertisements and supportive editorials, and that information tended to be vague. When Black Brothers embarked on an ambitious campaign in 1904 to nationalize their market for processed salted codfish,[26] their messages to Canadian retailers instructing them to be advocates of healthy eating often made generalized references to health experts: "Tell them that the opinion of the best authorities on diet is that people should EAT LESS MEAT and MORE FISH."[27] Companies like Black Brothers needed to link the healthfulness of fish directly to their specific brand names: "Teach them that fish is a healthier diet than meat, and teach them particularly that they ought to eat either 'Halifax,' 'Acadia,' or 'Bluenose' Brand Codfish."[28] This approach was more typical for canned or dried products than for fresh or frozen fish. By placing paid advertising in a national trade paper like the *Canadian Grocer*, Black Brothers and others also got supportive editorials from the paper. *Canadian Grocer* frequently implored grocers to expand their fish business and provided helpful articles on how to handle, store, and present seafood.[29] In some editorials, unidentified authors simply lifted language from Black Brothers' advertisements that emphasized "the value of fish as a nutritious diet."[30]

Black Brothers ads continued to applaud "good, healthful, nourishing food,"[31] claiming that their "brands of fish are both wholesome and good to eat,"[32] and "Good habits and 'Acadia Prepared Codfish will keep anyone healthy and happy."[33] Another popular trope was "Light, nourishing, appetizing sea food" for summer months.[34] In advertising its new line of Halifax Fish Cakes, Black Brothers insisted there was "nothing nicer, tastier, or more healthful" and fish cakes were a product consumers would continue to demand: "The reason lies in the permanency of the article. It isn't a fad or a fancy. It's just good wholesome, nourishing, tasty food that people will like to see on their tables quite frequently."[35] This particular advertisement had the irony of capturing the health-food fad of the early twentieth century while insisting it wasn't a fad. The food fads of the 1900s had been problematic for seafood producers as customers moved away from traditional foods like pickled and salted fish towards new processed and packaged foods like breakfast cereals and canned fruit and vegetables and meat. Black Brothers tried to challenge this by characterizing fish as traditional, wholesome food in contrast to "new-fangled food" and breakfast "fads."[36]

Some Black Brothers advertisements used more specific messages. Fish as important body and brain food was a common industry claim. Black Brothers asked grocers and jobbers to think more about fish as a potential breakfast food and push the "body-building, brain-strengthening kind [of breakfast] that is so popular in Nova Scotia." A fish breakfast helped prepare Martitimers for "a hard day's work."[37] Fish was "for the man who works both brain and hands,"[38] and Black Brothers provided "a beautiful white, tasty fish food that's easily digested and very stimulating both to body and brain. There is not another foodstuff at once so tasty and so good for the system."[39] Many seafood advertisers made the claim that fish protein was more easily digested. Given the high meat and starch diet of the typical Canadian during the early twentieth century, such assertions were probably welcomed.[40]

When Black Brothers subsided in the market by 1910, Connors Brothers emerged as the most aggressive advertiser of fish product, but its message remained focused on the modernity of its plants and purity of its product. Connors Brothers did occasionally reference health in the same vague language as Black Brothers before it. A 1918 Connors ad simply referred to the "wholesomeness" of seafood.[41] In 1920, the company claimed "The delicious, satisfying flavor of these wholesome sea foods wins the esteem of the most fastidious,"[42] and in 1924 simply stated that sardines were "so nourishing."[43] Connors used expert endorsement when opportunities emerged, such as when the Department of Marine and Fisheries released data on the nutritive value of seafood, placing canned sardines near the top in terms of calories per pound. Connors was quick to market this information, noting, "The superior quality of Brunswick Brand has been proven by Government analysis to be more nutritious and *less expensive* than imported lines,"[44] and "Brunswick Brand Sea Foods have been proven by a recent Government test to be superior in Quality and Food 'Calories' to any imported lines."[45] In fact, the government study simply said that sardines had high fat and calorie levels per pound but made no distinction between imported and domestic sardines. Yet Connors Brothers was in a marketing fight with imports and so combined data on the health value of sardines – which would be universal across all producers – with a call for nationalism in consumer purchasing. This line of marketing would have been attractive at a time when economic nationalism emerged as a powerful political force immediately following the war. Because the fish being marketed were caught by Canadian fishermen, processed by Canadian workers, and transported by Canadian shippers, the company claimed that any purchase of Connors' product was an addition to the economic value of the nation, "a quality which Canadians are proud to own."[46] Although the nationalist campaign began before the war, it became more common in company advertising once hostilities broke out and after, as Connors emphasized "those delicious sea foods are handled and processed entirely by Canadians."[47]

Industry leaders began to cultivate closer relationships with the federal government following their wartime experience with food rationing and marketing. The 1923–24 industry-government cooperative advertising was one of the first national seafood marketing campaigns involving more than one company or local cooperative. As such, government-funded marketing had to turn away from the industry focus on brand-name marketing that attempted to separate individual companies from the overall poor consumer image of seafood. Government-funded marketing had to directly combat that negative image and emphasize the quality and healthfulness of all Canadian fish and seafood products, while at the same time trying to make a distinction between imports and Canadian-produced products.

Alfred Herbert Brittain, general manager of the Maritime Fish Corporation and chair of the Publicity Committee for the Canadian Fisheries Association, noted the necessity of vagueness when referencing fish products in advertising because the organization could not place the value of any one fish species or product over another. The campaign therefore told "people generally how much better their health would be by including at least two meals a week as a fish diet."[48] One example of this advertising was the "Are You Fit? If Not Here Is the Secret!" spot (figure 2.2) by the Canadian Fisheries Association. This advertisement featured Bill Hughes, coach of Queen's University football program, suggesting that fish should be part of the "regular item on the weekly training programme of athletes." According to Hughes, fish was even better than meat because "its strength is assimilated by men in training with comparative rest to their digestive organs."[49]

Such references would have played well in the context of the era's celebration of athleticism. Sport historians have shown the rising importance of organized team sports, both professional and amateur, as it related to North America's masculine identity.[50] Yet as historians also know, such celebration of masculinity was a direct reaction to the reality that many men were not engaged in physical work and that an increasing amount of middle-class "labour" was deskwork. The rising popularity of organized sports, camping and hunting, masculine-themed dime novels, and even the Yukon Gold Rush have all been characterized as responses to monotonous desk labour.[51] Fisheries advertising reflected the new social reality. Another ad titled "If You Work 'Inside' Eat More Fish" (figure 2.3) presented fish as the solution to the "high blood pressure" and infamous "nerves" that plagued modern society.

Other advertisements had a more scientific focus on vitamin and mineral values. In its Eat Fish for Health campaign, the Department of Marine and Fisheries and the Canadian Fisheries Association emphasized the healthfulness of fish for all, but particularly for children.[52] They emphasized that fish contained high levels of vitamin D needed to build stronger bones and ward off rickets, while vitamin A in fish helped with eye health. In these advertisements, the expert was often an unidentified doctor who articulated in simple language the new advances in the science of nutrition that had recognized the importance of vitamins and minerals now seen as essential to health and strength. Again, the larger historical context is key here. Following the

ARE YOU FIT?
If not, here is the secret!

This is the advice of two of Canada's foremost athletes and trainers of *real men*.

"Bill" Hughes

The famous coach of Queen's University football team, the champions of the Dominion, 1922 and 1923.

Mr. Hughes says:

"I heartily subscribe to the beneficial results obtained by including fish as a regular item on the weekly training table programme of athletes."

"It has seemed to me that not only does Fish afford a welcome change from the ordinary diet of meats—but that its strength is assimilated by men in training with comparative rest to their digestive organs."

"This is no small factor in keeping from going stale under the pressure of hard work. After all, physical fitness means keeping in order every part of a man's physical make-up."

Dr. Donnelly says:

"From a medical standpoint, I would like to say that the sooner the country realizes what a valuable asset to the health of the community the consumption of more Fish and less meat would be, the better the general health of the community will be. I really think that many disease conditions prevalent in our communities are directly due to inefficient feeding of the individual, and a more regular use of Fish might help to control many incidental conditions of digestive disturbances now so prevalent amongst our people."

Dr. "Joe" Donnelly

Honorary Coach of Loyola College, who has guided the team in winning the Quebec Intercollegiate, the Dominion Intercollegiate titles and also the Quebec Interprovincial and Dominion Interprovincial titles during 1923.

Eat More Fish!

Fig. 2.2   "Are You Fit? If Not, Here Is the Secret!," Canada Eat More Fish campaign by the Canadian Co-operative Fish Publicity Fund, July 1923 to July 1924. The government- and industry-funded Eat More Fish campaign of the mid-1920s tapped into the changing sense of masculinity of the late industrial period. This advertisement linked seafood consumption to "real men" and the popularity of college football.

First World War, North Americans became obsessed with fears of malnourishment and a "vitamin and mineral famine."[53] This played particularly well with women in their role as housewives and primarily responsible for their children's health.

If you work "inside"

Eat more fish

There's FISH to stuff
    and FISH to bake;
There's FISH to broil
    just like a steak;
There's FISH for chowder
    and to fry;
Or any way you
    care to try

EAT MORE FISH

Tasty Fish Recipes

Educational Division,
Canadian Fisheries Association,
P.O. Box 1934, Montreal.
Please send me, free and postpaid, a copy
of your COOK BOOK, containing 69 Recipes for Cooking Fish.

Name

Address

If you have high Blood Pressure
Eat more Fish
If you are getting fat
Eat more Fish
If you are developing "nerves"
Eat more Fish

Easy of digestion
—no heavy feeling after
eating fish

Fig. 2.3   "If You Work 'Inside' Eat More Fish," Canada Eat More Fish campaign by the Canadian Co-operative Fish Publicity Fund, July 1923 to July 1924. As more men found themselves confined to white-collar desk work, advertising that suggested improved vigour via diet would have captured attention. This ad highlighted health concerns around blood pressure, weight control, and neurasthenia.

Although the theme was common in other food commodity advertising, promotion of Canada's seafood needed to emphasize something uniquely naturally valuable about Canada and its seafood resources over imported seafood. To do so, advertisers argued that the uniquely pure Canadian environment made Canadian seafood products higher in nutritious quality (figure 2.4). Such ads celebrated the oceanic source of the product, pointing rather vaguely to the vastness and purity of the ocean and the inexhaustibility of the resource. This message created seafood consumers increasingly divorced from the reality of resource depletion who were thus shocked in the second half of the twentieth century by newly voiced concerns about degradation of the world's oceans.

"Eat Fish for Health"

Fish comes to your table
from the cool, clear depths
of Canada's seas and lakes
and rivers — clean, firm-
bodied, wholesome fish —
good to look at and good
to eat because rich in the
very food elements your
system needs.

Healthful, nourishing and tasty,
fish brings you a new eating
pleasure. Fish brings you the
little spice of variety, the little
change from the same old meat
dishes, the change that appeals
to the eye and tempts the
appetite.

A free cook book showing how
to prepare fish in scores of delightful ways will be sent you if
you write—

Educational Division
**CANADIAN FISHERIES
ASSOCIATION**
Board of Trade Building
MONTREAL

59

Fig. 2.4   "Eat Fish for Health," Eat Fish for Health campaign by the Canadian Co-operative Fish Publicity Fund to Increase Demand for Fish, July 1924 to July 1925. The Canadian Fisheries Association partnered with the federal government in the 1920s to encourage increased seafood consumption to improve industry well-being. To avoid emphasizing one branch of the diverse fisheries industry over any other, the advertisements presented vague, generalized images of fish that failed to expand consumer knowledge.

Environmental historians have documented the utilization of environmentally themed marketing strategies, the cultural impact of such strategies, and the implications of large-scale, highly concentrated consumer-base food production for resources. Environmental histories of California and Florida fruits and vegetables have been particularly abundant.[54] Seafood, however, has failed to attract the same level of interest among either environmental or food historians, nor, surprisingly, have fisheries or maritime historians paid much attention to the consumer history of seafood.

An examination of seafood marketing in Canada reveals how closely related it was to larger celebratory rhetoric of Canada's northern culture. Early in the twentieth century, Canadian culture began to celebrate its northern identity, often associating the cold vastness of its geography with the assumed purity of its people. Seafood marketers were aware of this larger context and sought to tap into "northern purity" ideals. In this context, of course, the purity of coldness was less associated with the snow- and ice-covered North than with the cool, clear ocean and coastal environments. For example, in 1904 Black Brothers of Halifax noted the superiority of "the air of the sea coast," which was "better and more invigorating than the air of inland parts." Inland consumers trapped in the modern burgeoning metropolis could access this natural experience so celebrated in the nation's popular culture by consuming it in the form of the "tasty cod of the Atlantic," which was "all the tastier when cured" by a salting process that used "just salt enough to be nice."[55] Likewise, Connors Brothers in a 1917 advertisement extolled its canned sardines as "fresh and clean as a breeze from the sea"; in 1918, it proclaimed them as "delicious as a breeze from the Old Atlantic."[56] In 1922, it continued to emphasize the natural "wholesomeness" of products extracted from the "clear cool depths" of the oceans, claiming, "nothing is left undone" to ensure the preservation of that naturalness throughout the processing of the fish into a food "ready to serve on opening the tin."[57] Such direct consumer access to the purity of Canada's nature via eating lightly salted codfish or the simple act of opening a tin worked well within the larger cultural context of celebration of the nation's rugged beauty.[58]

The Canadian federal government also tapped into this idealization. In 1930s radio addresses, fisheries scientist Mildred Helena Campbell proclaimed, "It is an axiom among authorities that food fishes improve the quality in proportion to the purity and coldness of the waters in which they are taken. Judged according to this fact, Canadian fish are unexcelled in quality anywhere in the world."[59] One of the government's earliest cook booklets also emphasized the natural superiority of Canadian fish products. The text of the Department of Fisheries' "Food and Health Value of Fish" (1930) emphasized the superiority of Canadian fish due to "purity and coldness of the waters in which they are taken." National propaganda tried to link this natural purity to consumers' health by emphasizing the high levels of protein, amino acids, and fat-soluble vitamins in specifically Canadian fish that gave it abundant food value. Furthermore, fish contained high levels of calcium for bone health and iodine to prevent goiter.[60] Of course, all sea fish have similar levels of health-giving elements, but by preceding the discussion on health with nationalist rhetoric of nature, the propaganda tended to direct the consumer to assume Canadian fish were superior to those of their southern neighbour.

Domestic supplies of fish, insisted H.F.S. (Herbert Frater Starr) Paisley,[61] director of the Fisheries Intelligence and Publicity Division within the Department of Marine and Fisheries, were "especially valuable foods" in comparison with imports.[62] Obviously, the federal government, necessarily interested in promoting the national economy, needed to separate Canadian fish from foreign competitors. Evelene Spencer, the government-funded "fish evangelist," also used this nationalism in her 1932 radio addresses, in which she celebrated the superiority of Canadian codfish over those of the United States because "the [Canadian] environment is so perfectly suited to their needs." In the North Atlantic, the cold current of northern waters collided with warm air of the Gulf Stream, Spencer explained, and the "net product is the finest codfish in the world!"[63] (Her addresses did not note that much of the fish imported from the United States was caught in the same Canadian waters as Canadian-marketed fish.)

Nationalism and nature were thus deeply interwoven in the marketing of the healthy qualities of Canada's seafood to its own domestic market. Advertisements argued that Canadian fish producers could more easily deliver the purity of ocean to Canadian consumers because of the close proximity of the environment to the processing of the food and to consumers. Processed seafood could metaphorically bring consumers right to the waters' edge. In 1915, Connors Brothers linked the superiority of Canada's sardines over foreign imports to this proximity: "Located close to the fishing grounds we get the pick of the fishermen's catches."[64] In 1916, the company claimed its factory was "close to what is probably the best fishing ground on the Atlantic" and so able to "eliminate all but the very choicest fish."[65] Arthur P. Tippet & Company of Montreal, wholesale distributor for Thistle Brand processed fish, also made vague references to the "best waters" or "Canadian waters" from which it acquired its seafood. A 1915 ad got a bit more specific about the source of the "choicest and best fish" as "the famous Nova Scotian fishing beds." The end result: "Thistle Brand Fish reach the table wholesome and appetizing."[66] On the Pacific Coast, Canadian Canners Ltd launched its Fresh from the Sea advertising campaign in 1936, noting that the oysters in its Cream of Oyster Soup came from the majestic Crescent Beach "only a few miles from our factory in Vancouver." The "rich dairy from the famous Fraser Valley" gave an "added touch of luxury" to the product, combining the purity of both land and sea.[67]

The common trope of celebration of nature and nation in the early twentieth century elevated Canadian national identity at a time when such nationalism was at the center of political debate swirling around the 1931 Statute of Westminster. Like that of United States and Great Britain, Canada's nationalism in this period revolved around work and masculinity, both easily combined with a celebration of nature and nationalism. Men in Canada had historically conquered nature to build a nation – or so the old trope went. This masculinity of work within nature appeared in seafood marketing as in other venues of popular culture. The American Can Company of Vancouver said eloquently in their marketing of canned clams and lobster: "Right at the shore where the rugged, hardy fishermen bring in their catches, the choicest fish, clams and lobster are quickly cleaned, prepared, and sealed in cans while their cold, crisp ocean goodness and freshness are at their peak."[68] More so than most, this ad copy combined key elements of nature idealization in Canada: its ruggedness, coldness, and purity, all part of Canada's northern – and masculine – identity.

Generally, however, the environmental themes in the marketing of Canada's seafood were vague and non-specific. This was particularly true in the government-funded marketing of the 1930s. To avoid accusations of bias, the government needed to market non-industry specific seafood; its images of the environment had to project a universalism. Thus, consumers were not educated about specific fish species, their environment, or the commodities they became. The fish in these advertisements was a fictitious species – a little bit salmon, a little bit codfish, a little bit herring, but exclusively none of them (figure 2.5).

Fig. 2.5 "A Crusade That Will Multiply Every Canadian Dealer's Profits," *Prairie Grocer and Provisioner*, September 1936, 27. The 1930s Any Day a Fish Day campaign of the Department of Fisheries targeted retailers as well as consumers, hoping to improve retail dealers' ability to purchase, store, and sell seafood. This 1936 ad in *Prairie Grocer and Provisioner* promised increased profits for retailers who offered more seafood to customers. It also reflected the government effort to expand the seafood market to the Prairies.

As such, even as the marketing of seafood during the 1920s and '30s emphasized Canada's superior environment, the Canadian consumer was not presented with specific references to that environment but was instead told of the general grandeur, abundance, and inexhaustibility of Canada's marine resources – even as policy-makers and scientists knew that those were in fact under threat.

By the 1930s, the generalized references to the wholesomeness of fish as food expanded to include more specific references to vitamins and minerals found in seafood and more utilization of expert medical and scientific data. This new, enriched language, much of it directed by the federal government through the Department of Fisheries, built upon earlier platforms on the value of fish as a "brain food," a source of protein to build "brawn," and an easily digestible food to help those who found themselves behind a desk. Post-1930 advertisements embraced new findings in food science and the growing knowledge about the links between food and health, specifically in relation to deficiency diseases. The American Can Company was an early advocate of using food science to market canned foods. Though not a seafood producer, it was directly interested in seeing increased sales of canned seafood. In a 1934 campaign, the company referenced "medical authorities" and noted that "not only is [canned salmon] one of the cheapest of all foods – it is one of the most nourishing, one of the richest in health-giving qualities." [69] Unlike earlier marketing efforts, this advertisement went further to define "health-giving" as vitamin D, the "sunshine vitamin," and vitamin A, and iodine, which "prevents that disfiguring disease called goiter." [70] In another ad, the company noted that canned seafoods "are rich in natural iodine, and the canned fish is rich in body-building proteins and Vitamin D." [71] In 1937, the company built upon earlier ideas that seafood was a magical boost to brain power in an imagined conversation between two average Canadians. "What do you mean when you say sea food is brain food?" one asks; the other replies, "It's not only good to eat but rich in vitamins and iodine ... so it's good *sense* to eat it often." [72]

Protein was another focus of this more specific health-food campaign. Ann Adam, in a 1932 column for *Canadian Home Journal*, noted that canned pink salmon had high "protein value" even if the flavour was less rich than the higher grades of salmon. Within the context of the Depression, Adam, like others, increasingly focused on food value – that is, nutrition as related to cost – and noted that even chum salmon, although not as rich in fats, still had the same "body building qualities [protein] of the more expensive varieties."[73] In a 1936 column, she described fish as "marvellously rich in vitamins and in minerals, the builders and protectors of health that we must seek to include richly in the foods we select."[74] Fish's high iodine and vitamin D content was also widely cited in advertising as a means to develop good bone health and prevent rickets. Another *Canadian Home Journal* article of 1932 explained that cod liver oil contained very high levels of vitamin D because the natural fats of the fish were practically all concentrated in the liver, and cod liver oil was especially important during winter months when people developed vitamin D deficiencies.[75]

Throughout the 1930s, after marketers got more specific in their references to the health qualities of food, advertisements regularly referred to experts. Although scientists or doctors were rarely mentioned by name, advertisements called upon the consumers to place blind trust in vague, generic references to superior knowledge. Food health was increasingly seen as a complex scientific matter beyond the average individual's understanding. Instead of trying to build knowledge, marketers instead asked consumers to put faith in food-health professionals. Such experts could be found in universities or government laboratories, but more often they were employed by the food industry, which, according to many marketing messages, had altruistic goals of providing consumers with the healthiest food options available. Media outlets dependent upon advertising had an obvious self-interest in perpetuating this blind-trust agenda. One 1932 editorial in the *Canadian Home Journal*, for example, encouraged mothers to have confidence in food industry ads when selecting the right foods for their children because "the leading manufacturers of food now seek the advice and approval of the leading nutritional authorities in the country, relative to the claims they make for their product in their advertising." The article continued, "Food advertisements are reliable sources of information regarding diet, based upon the results of the latest approved scientific discoveries about vitamins, minerals and roughage."[76] In a subsequent piece, the magazine stated that advertisements were not just about selling products but were in fact "educational" because "scientific minds contribute to their contents. Their recommendations are based on deep thoughts. Their words carefully chosen; their dictation studiously formed for clarity and undertaking." Through ads, mothers could learn from respected authorities about children's health and decide the best products to buy based on advanced knowledge. Consequently, it was always best "to buy the advertised brand."[77]

Not surprisingly, the industry also turned to the expert image of the federal government to build consumer confidence in the product. In 1930 the Department of Marine and Fisheries funded a radio campaign that featured a young biologist, Mildred Campbell. Campbell had earned a BA (1926) and an MA (1928) in zoology under the tutelage of Dr Charles McLean Fraser, from the University of British Columbia, where she had aided Fraser's work on copepods parasitic on fish. She was only the second woman to receive an MA from UBC and went on to be an instructor in biology there in 1929 and 1930. She also received a National Research Council grant in 1929 and earned a PhD from the University of Toronto in 1933, again with a focus on copepods. She went on to do research in St Andrews, New Brunswick, and Woods Hole, Massachusetts. Campbell published five academic papers before her professional career was brought to an end by marriage. In 1934, she married Wilbert Lloyd Attridge, a dentist, and settled into the role of wife and mother in Toronto, never to return to science.[78] The contemporary social assumptions of a woman's role limited this promising young scientist's career opportunities within her profession, and it appears that the male-dominated Department of Marine and Fisheries saw her talents as useful only to advise women about cooking fish.

Campbell's radio talks highlighted what had become the three key health components of fish: protein, easily digestible, rich in vitamins (specifically vitamin D). "Protein content is the chief food constituent of fish and contains all the essential amino-acids from which human muscle and many other tissues are constructed," Campbell emphasized. She wove into her talk numerous references to medical and scientific experts, inferring a trust in authorities on the part of the consumer. She referenced Dr Donald K. Tressler of the Mellon Institute of Industrial Research, who articulated the importance of fat-soluble vitamins and cod liver oil as well as the importance of iodine. She discussed Dr D.C. Hall's research on iodine-rich diets of the American Northwest, and Dr Elmer Verner MacCollon's studies on the dangers of rickets and bone defects among children, which were rare where people regularly consumed fish.[79] Campbell did not go into detail on medical and scientific studies but instead relied on statements about the health qualities of fish, specifically their ability to ward of deficiency diseases.

H.F.S. Paisley noted in his radio address for the Institute of the Civil Service in March 1930 that "scientists [had] established that fish are especially valuable foods, because of their richness in vitamins and iodine and other elements."[80] Paisley quoted many of the same authorities as Campbell, adding that Dr John A. Amyot, Canada's deputy minister of national health, had specifically noted the high content of vitamin D in fish as necessary for bone health. Paisley also incorporated the work of Dr Alton Goldbloom, a lecturer in the Faculty of Medicine at McGill University, and Dr Elmer McCollum of Johns Hopkins, both of whom again noted that rickets was rare in populations that consumed fish. In 1932, William Found, deputy minister of fisheries, also noted that "medical science" had determined that fish was a "highly palatable as well as an unusually nutritive and economic food."[81]

In most of her talks, however, Campbell avoided specific references to experts. Instead she simply highlighted that "fish are not only nourishing and readily digested, but they are abounding in elements which are health-guarding." She reiterated that fish was an inexpensive source of nutrition and particularly important for young children to provide the "very materials that are needed to enable them to grow healthy and strong." Campbell also highlighted the importance of fish for a stressed Canadian population struggling with the realities of the Depression. Unlike heavy meats, she maintained, fish was easily digested, "a consideration worth remembering since there is so much nervous tension in life nowadays."[82]

The context of the Depression led Campbell and the Department of Marine and Fisheries to address the economic realities of purchasing power as well. In one presentation, Campbell noted, "Generally speaking, Canadian fish is cheaper than other foodstuffs of comparable food value [nutrition]. It is as rich as meat in nutrients, and in some cases it is richer in nourishment." With dual consideration of family finances and family health, she assured her female audience that fish was the best and most "economical source of energy."[83] Her radio talks, lasting from 1 April until 24 June 1930, each opened with some reference to the health qualities of fish. She also occasionally referred to the "rigorous" government inspection system that ensured high quality, the variety of fish products available to diversify the family menu, the convenience of canned seafood for quick summer meals on the go, and the value of incorporating fish into each day's meal plan. Newspapers carried some of her talks in full, and Campbell attended public functions at department stores and cooking demonstrations.

The department seemed satisfied with the impact of the radio and public demonstrations. In 1932 it employed Evelene Spencer, who had already conducted a nationwide speaking tour in the United States for the US Fish Bureau.[84] Spencer began nearly every one of her presentations, at live demonstrations or over the radio, with what became her catchphrase: "I find that men love fish but women hate to cook it." Her stated aim was to make Canadian housewives "fish conscious," to impress upon them the nutritive and economic value of adding more fish to the family diet and to instruct them on the best ways to select, prepare, and serve fish meals. In Ottawa in February 1932, she applauded how – in a line she would often repeat – "the great, big Canadian public have become aware of the vitamin need in their food." Fish was central to this rising awareness of food health because of the important vitamins and minerals it provided, especially vitamin D for warding off the dreaded rickets in young Canadian children: "This is most important as we all want to see our children grow rugged, sturdy legs and bodies. There is no more pathetic sight in the world than that of a feeble, weak boned child with badly bowed legs."[85] Yet such references to health benefits were rare in Spencer's talks; instead, she largely focused on cooking methods. The rhetorical themes of her presentations fit more clearly in the age's constructs of gender roles; that as housewives, women were responsible for providing healthy, varied, and cost-effective meals for their family. She deployed the dominant gender themes of her day to convince women to use seafood in meal planning. (Spencer is discussed in more detail in the next chapter, which deals specifically with gendered rhetoric in seafood marketing.)

Campbell's and Spencer's broadcasts were not a new strategy for the Department of Marine and Fisheries. Throughout the 1930s, it regularly used the radio to market to Canadians, often emphasizing its scientific work and enforcement of conservation and inspection laws.[86] Other than Campbell's and Spencer's, however, only a few departmental radio addresses focused on fish as food. Basil E. Bailey of the Pacific Fisheries Experimental Station defended the idea that, although the effects of nutritional deficiencies of humans belonged to the field of medical and agricultural research, fisheries biologists too should undertake research on the nutritional properties of fish products. Unlike experts in the medical and nutritional fields, Bailey argued, fisheries scientists recognized the essential ecological factors that made fish valuable food. An ecological approach showed that the same minerals needed for the human body were concentrated in ocean water; thus fish "live and breathe an atmosphere which contain many of the known elements" essential for human health.[87] Similarly, Dr Wilbert A. Clemens, director of the Biological Station in Nanaimo, BC, presented a talk titled "The Sea as Nature's Storehouse of Food Materials" in December 1934. Clemens recognized the popularity of beef and other terrestrial meats but argued that "recent advances in scientific research in the fields of food chemistry and nutrition have shown that sea foods are pre-eminent in food values." After going into some detail about the chemical makeup of the human body and the need to ingest a variety of important minerals and vitamins to maintain health, Clemens noted, "Strangely enough, the sea is a storehouse of nearly all of these materials which go to make up our bodies."[88]

In their talks, Bailey and Clemens, unlike Campbell and Spencer, were detailed in their scientific explanations of oceanography, marine biology, and human chemistry. They made environmental connections between human biochemical needs and the rich mineral content of ocean water, plants, and animals. Clemens noted, for example, that animals (including humans) rely on plants for survival; but whereas minerals in soils are processed slowly through weathering and solution, minerals of the oceans are already in solution form and thus more easily available for marine plant growth. Mineral-rich seas produce mineral-rich sea plants, which are then consumed by mineral-rich sea fish, which become mineral-rich seafood.[89] Clemens made specific use of data on sea salts and iodine as preventatives for goiter and suggested that "'Fish Friday' might be termed 'iodine Friday.'" He provided similar in-depth analysis of proteins, fats, oils, and vitamins in calculating the superiority of seafood for human health. Fish fat, he claimed, was more easily converted into heat by the human body than other fats and oils. In terms of the "mysterious class of chemical substances known as Vitamins," he contended that seafood provided high levels of vitamin A for eye health, vitamin B to promote body growth, vitamin C to prevent scurvy, and vitamin D for bone and teeth.[90] He called for a more informed consumption of food, arguing that anyone who fully understood the biochemistry of fish and food would consume the oil from canned salmon rather than draining it off. Bailey too warned Canadian housewives not to be dismissive of pink or white spring salmon just because it did not look as nice as red spring salmon; the "thrifty housewife will do well to remember that the colour is not necessarily an index of food value." Red, pink, and white contained relatively the same levels of protein and vitamins A and D, and "the red pigment itself has no dietary significance." Their talks were unusual in their detailed scientific analysis, but Bailey and Clemens were hardly out of the ordinary in their overall message: eat fish for health.

These radio addresses immediately preceded a massive government advertising campaign that began in 1936 and continued until 1940. The main rationale was largely one of rehabilitating the fisheries industry at a time of economic crisis (discussed in more detail in chapter 4). Yet communicating the idea that fish could contribute to both national health and individual well-being at a time of nutritional crisis also received the Department of Fisheries' attention. In 1937, Minister Joseph-Enoil Michaud[91] articulated a similar argument in a radio broadcast from Halifax: "Perhaps the reason [for low consumption rates in Canada] is that Canadians have not all realized that there are scientific reasons why our fish and shellfish make particularly good foods for growing children and adults alike."[92] Expert fisheries scientists employed by the department could thus provide a valuable service to the nation by overseeing the marketing of fish to ensure the focus was on the well-being, nutritionally and economically, of the nation. Instead of individual industry leaders pushing their specific products, the government campaign aimed at advancing consumption of fish and fish products as a public service. If the Canadian people collectively ate more fish, the nation would become a healthier society; thus, the government-funded campaign might be justified as a public health program.

While family economics, the appealing taste of fish, and the ease of preparation all had a part in advertising content, it was the health-giving qualities of fish that played the leading role. The main health themes of government advertising remained closely in line with those that had emerged in the early 1930s: fish was good for all people's health but particularly children's; it was an economical source of easily digestible proteins; and it had high levels of health-giving oils and fats, important minerals like iodine and copper, and high levels of vitamin D for good bone health. [93]

Most of the advertisements targeting women consumers mainly concerned children's health. In radio spots produced by the Atlantic

Fig. 2.6 "Radiantly Healthy," Department of Fisheries' Any Day a Fish Day campaign, October 1937. By the 1930s, as Canadian women were increasingly in control of household purchases, especially food, the Department of Fisheries' advertising campaigns targeted housewives' responsibility to provide wholesome family meals. Advertisements touted seafood as "nourishing food for active youngsters" that provided protein, minerals, and "protective" vitamins.

Fig. 2.7 "D'une santé radieuse!," Department of Fisheries' Any Day a Fish Day campaign, October 1937. The Department of Fisheries made little effort in its fish marketing to appeal to Canada's ethnic minorities, with the occasional exception of the Jewish community. Its French- and English-language advertisements were identical in content, with the same emphasis on health and women's role in providing for their families.

Fig. 2.8 "Strong Bodies, Sound Teeth, Good Growth, Promoted by Fish," *Hamilton Spectator*, 21 October 1936, 13. According to the Department of Fisheries, there was "no better food" for growing bodies than fish. Marketing campaigns increasingly focused on women's role in providing healthy foods for families. Advertising regularly noted that seafood contained "protein for strength and energy," vitamins for disease resistance, calcium and phosphorus for bones and teeth, and copper for "good blood." Although some advertisements included images of girls, active boys were much more commonly shown.

Advertising Agency for February 1937, for example, the Department of Fisheries told listeners in the Maritime provinces that fish were "ideal foods for supplying the protein requirements in the diet of growing children, without the stimulation that meats afford."[94] In a 1937 ad that circulated nationally, the Department of Fisheries advised giving young boys "plenty of good Canadian fish at meal-times" so they would "grow sturdy and strong, as healthy as a horse."[95] Other advertisements made more generalized references to "the vitamin fish" or "nourishing" elements.[96] A 1938 ad (figure 2.6) claimed fish gave "grand nourishing food for active youngsters always 'on the go.' It peps them up, and restores their energy."[97] A 1939 message in the French-language *La Revue Populaire* (figure 2.7) stated that fish was the ideal food for schoolchildren.[98] The "romping roughhouse" of play required "plenty of fuel to give flame to youth," another ad series that year noted; fish provided "the nourishment and energy that Nature demands."[99] Many advertisements featured young children – usually boys – playing (figure 2.8), and called upon mothers to provide "fish for family health," emphasizing that fish diets rich in vitamins A and D, phosphorus, calcium, iodine, and copper promoted "strong bodies, sound teeth, and good growth."[100]

Although mainly focused on health benefits, some marketing was tailored to regional, gender, and ethnic-specific audiences. To farmers in the Prairie provinces, the Department of Fisheries promoted fish as "Nourishing, Strength-Building Food for Farm Meals." These ads, often featuring tractors plowing fields, cited the protein-rich content of fish that "builds up strength, promotes physical well-being."[101] Although not common, some advertisements targeted men. These featured more masculine themes like physical fitness, "pep," or that "rare old 'up-and-at-'em' feeling," and advised men to "ask your wife to make more fish."[102] In Jewish papers, fish was identified as "the food of tradition": "its wholesomeness and delicate purity makes it always the appropriate dish for ritualistic observances a well as everyday meals."[103] Department ads in French-language papers and magazines carried the same general "Any Day a Fish Day" message as their English-language counterparts: "Du poisson n'importe quel jour."[104]

Magazines and newspapers often provided editorials supportive of the paid advertising. Katherine Caldwell, of the Home Service Bureau of the *Canadian Home Journal*, was a regular writer of such editorials. In one, she championed increased fish consumption around the Great Lakes, which she referred to as the "goitre belt" of Canada. Fish's iodine content, she claimed, would quickly relieve the population from the trauma of this deficiency.[105] Jean Firth of the *Star Weekly*, writing under the pen name Jean Brodie, provided similar-themed editorials supporting the department's advertising.[106]

Minster of Fisheries Michaud added his signature to each ad, adding an authoritative presence that carried more weight than marketing ploys by private-sector industry. Coming from the voice of the fisheries within the national government, the advertisements (perhaps better identified as propaganda) conveyed a sense of public good in a publicly funded campaign dedicated to advancing Canada's health. In an open letter in the nation's newspapers, Michaud told Canadians that scientists had discovered that "certain mysterious substances now known as vitamins are essential to human health," and that "fish are rich sources of the chief of these health-builders." Fish protein was easily digested, and a wide variety of seafood was available to all Canadian consumers at all times of the year. Such messages from a government authority appeared as altruistic, objective messages intended to benefit the consumer, not industry growth. "Fish foods are health-builders and health-protectors," Michaud concluded. Increased national consumption would promote national public health.[107]

As the government advertising dramatically increased, industry advertising in trade papers like *Canadian Grocer* and national magazines like *Canadian Home Journal* significantly declined. Some seafood producers tried to capture the reality of an economically depressed market by pitching seafood as providing the best food value, with the most protein and nutrients per unit cost. An American Can Company ad featured a dutiful housewife responding to her husband's request for more canned seafood: "Suits me fine, keeps *me* slim and my *purse* fat. It's very economical you know."[108] (Interestingly, fish were marketed as a source for muscle gain for men, yet a food that was slimming for women.) In another ad, a gleeful wife and mother announced on seeing her husband and son eagerly eating up a canned fish meal, "Would you believe it – that meal was the most inexpensive one of the whole week."[109]

For the most part, however, industry advertising was limited to the trade press rather than direct consumer marketing and continued to ask grocers and jobbers to push their product to the customer. Connors Brothers proclaimed, "Brunswick Brand is the greatest sardine value that you can offer your customers ... sold at a price that amazes the average individual."[110] Their line of products provided "sardines for every price class."[111] In trade papers, companies emphasized their product's ease of sales: it "pays to display"; sardines were "quick sales" and "no shelf warmers."[112] Some industry leaders did seek to supplement the government's Any Day a Fish Day campaign with their own brand marketing, but, for the most part, government advertising presented consumers with non-label-specific seafood and general references to the health value of fish products.[113]

The continued insistence by the Department of Fisheries that fish was the best food for Canadians did not go without criticism. In 1939 the Department of Pensions and Public Health warned the Department of Fisheries that their use of the word "health" might violate Section 7(e) of the Food and Drug Act. Then in January 1940, Robert E. Wodehouse, deputy minister of pensions and public health, wrote to John J. Cowie, acting deputy minister of fisheries, including a clipping of an advertisement that department had placed in the *Evening Citizen*, noting that the references to "health food" violated the Food and Drug Act in that they were "misleading and exaggerated." While fish might be a good food, Wodehouse wrote, "it cannot be considered in the nature of a health food." Furthermore, the Department of Pensions and Public Health did not permit any food product to be marketed as a "health food." Wodehouse also disputed the idea that fish could build "family health," promoted "physical fitness," was "rich" in vitamins and mineral salts, and was a special health-giving food above and beyond other foods. Instead, he maintained, fish should be seen as one food choice among many, with nutritional value similar to other quality foods on the market.[114]

Cowie responded that such information had already been communicated to their advertising firm, Walsh Advertising Company, and that the new campaign would change its focus.[115] Indeed, in early 1939, Paisley instructed Walsh Advertising to remove from its ads the slogans "Fish – A Great Health Food" and "One of Nature's Greatest Health Food" and replace them with "Fish – Food for the Family."[116] The focus of the department's 1939–40 campaign shifted away from fish as a health food to the Canadian housewife utilizing fish to vary the family diet.[117] This change did not happen fast enough for Wodehouse. In February 1940, he again wrote to Cowie, calling the advertisements a continuing "embarrassment" to the Department of Pensions and Public Health.[118]

Whether marketing efforts to popularize fish as a health food were presented as vague references to health and wholesomeness, as between 1900 and about 1930, or scientifically and medically, as exemplified by Dr Clemens of the Biological Board of Canada during the 1930s, all fit within the general context of the early twentieth century, when the public became concerned about nutritional health and the potentially debilitating effects of deficiency diseases. Historians today may look back on Clemens's suggestion that Canadians consume more fish to acquire vitamin C and ward off scurvy as a bit irrational, as scurvy cases were by then rare in Canada. Yet such dismissiveness would fail to recognize the broader context. People like Clemens, Spencer, Campbell, and Michaud, as well as marketers for the fisheries industry, accepted the science of those they considered experts – medical researchers like Frederick Tisdall and Lionel Pett, who worked throughout the 1930s to convince people that the hidden hunger of malnutrition existed and that the government and the food industry must partner up to address it. Indeed, from 1936 to 1940, the joint campaign of the Department of Fisheries and the Canadian Fisheries Association sought to achieve the goals of medical and nutritional experts: improving the nutritive health of Canadians while also achieving a more industry-specific objective of economic recovery. The end result was an onslaught of advertising designed to convince Canadians that they should eat healthier food, for the sake not just of their own health but of that of the nation, and that fish should play a greater part in that healthier diet. Historical context suggests that such efforts were grounded in real concerns about nutritional deficiency in Canada and a belief that fish could play a role in preventing it.

3

Recasting the Seafood Consumer:
The Housewife and the Modern Kitchen

Recasting the image of fish was part of a larger change in food economics that emerged by the end of the nineteenth century as brand-named products began direct-consumer marketing campaigns in an effort to develop buyer loyalty to a particular brand and a surge of new convenient and processed foods entered the marketplace. The marketing of seafood between 1900 and the 1940s promoted fish and shellfish as modern foods, competitive with other modern and processed food products. To do so, the industry had to shed the historic association of seafood with the low quality and the unpalatability of salted, dried, and pickled fish. A key tool used in recasting seafood as a modern product was the emphasis on its health-giving qualities, as the previous chapter discussed. To further break the association with poor food, or food for the poor, the industry also crafted a message of modernity of the seafood consumer. Central to this marketing was an image of the middle-class, white Canadian woman (figure 3.1).

Just as the rhetoric of health redefined the image of the seafood product, the rhetoric of gender and class redefined the images of the seafood consumer in the government and industry's marketing. Seafood marketing in the early twentieth century was part of a larger culture that framed Canada as a nation of white, middle-class families with a domestic space controlled by the wife and mother.[1] The continual emphasis on the housewife as consumer, and the sometime diametrically opposed imagery of the masculinity of seafood production, did not take account of the role that Canadian women played in production, especially in the canneries on the Atlantic and Pacific coasts where women dominated the ranks of processing

Fig. 3.1 "Eat Fish for Health," edited proof, January 1937. Although family health was the main theme of the Department of Fisheries' 1930s Any Day a Fish Day campaign, marketing also directed consumers' attention to the economic value for women in serving more fish. Advertisements often included recipes and invited consumers to request a free fish cook booklet.

labourers. Just as the government and industry invested Canadian female consumers with agency through purchasing power, they de-emphasized their active role as producers. Further, the marketing emphasized middle-class purchasers and discounted (or tried active-ly to erase) the industry's historic reliance on low-income consumers of pickled, dried, and salted fish.

By projecting images of femininity and class, seafood marketers also participated in the construction of a Canadian national identity. Because the fisheries were so central to Canadian national economy, and, more importantly, Canadian national heritage, the marketing implied, consumers should develop buyer loyalty to domestic-sourced products. Canadian housewives, as the primary shapers of family food selection, not only shaped their families' health but as a collective whole also shaped the health of the nation. Public heath thus depended upon informed women making good nutritional choices. Yet this pro-jection of national health was not broadcast across all class lines. Gen-dered seafood marketing celebrated the wife and mother as the core of an emerging middle class idealized as the basis of a new national identity. Moreover, gender and seafood nationalism went beyond national public health and middle-class identity to include the bold idea that Canada's women, via their consumer spending power, held the key for industrial stability and recovery. If women bought fish, the final part of the argument went, they would rescue an important sector of the national economy from economic ruin. The fisher-ies would recover, Atlantic Canada would rebound, and the nation would survive.

Food thus offers a unique entry point into the history of Canada's homes and the relationship between home and nation. It is a conduit into the domestic life of the past. Via food, historians can come to some understanding of domestic values and the roles played by moth-er, father, and child within a household. In 2001, food historian Sherie A. Inness articulated food history's value as a method of historical investigation: "The foods we eat or do not eat, who prepares them, and how they are served reveal a tremendous amount of information about how a society is structured; food is one of the most visible and omnipresent symbols of everything from class to race to age, and it provides a powerful symbolic message of who we are and who we aspire to be." Although all people consume food, food holds much more than just physiological value; it is also deeply cultural. Inness also wrote, "Every society invests symbolic importance to food."[2]

Between 1900 and 1950, however, the Canadian Department of Fisheries and the Canadian Fisheries Association invested significantly in creating a *new* symbolic importance for fish. Their combined efforts sought to recreate the established class stereotype of fish as a food of the masses and to tie it instead to middle-class consumers. They did so largely via the power of women as middle-class cultural agents, with the aim of using women as a bridge into the middle-class diet as an expanding market for seafood sales.

Food history and women's history have long been intertwined. Historians have identified the roles women played in food selection and preparation as both empowering and limiting. Although culinary ideals often trapped women into a narrowly defined sense of self, and the marketing of an ideal kitchen limited the acceptable roles women could play, the power that society often gave to food and the authority women had over food selection provided women with an avenue of strength in defining the family's societal boundaries. They could be both active shapers of this process and passive consumers of values established by other forces, such as industry, government, or academia. Patricia M. Gantt noted in her essay on American southern food, "Foods reveal what signals a culture is sending, as well as what use women make of those signals. Of notable importance is whether women accept these cultural codes at face value, internalizing messages sent as guides for their behavior, or reconfiguring them to their own design."[3] Canadian fisheries certainly tried to send messages to Canada's housewives to redefine their behaviour by making them seafood shoppers and consumers. Considering the minimal increases in sales as a result of this decades-long campaign, we might conclude that Canadian women rejected this messaging and retained control of their kitchens via the choices they made.

This study of Canadian seafood marketing reflects four key themes at the intersection of food history and women's history: knowledge of food via advertisements and commercial cookbooks; the increasing cultural relevance of the middle class; consumer economics; and the rise of professional home economics. Accessing knowledge about food via industry advertisements and commercial cookbooks represented a significant generational break at the turn of the century as many women turned away from their mothers and towards industry for guidance in food selection and preparation. Jessamyn Neuhaus notes in her informative essay "Is Meatloaf for Men? Gender and Meatloaf Recipes, 1920–1960" that "the post-World War I boom in cookbook publication coincided with a marked decline in the number of household servants in middle-class homes. As more married middle-class women faced the kitchen for the first time, often without the support of an extended family structure, more women turned to published cookery advice."[4] The Canadian fisheries tapped into this insecurity and sought to instruct new, white, middle-class wives and mothers on how to be good cooks, incorporating even difficult-to-work-with food products (such as seafood) without the traditional help of servants and extended family members. But cookbook publishing was not an isolated industry. It worked together with the food industry and, often, with the government agencies most dedicated to promoting the food-production industries. The cookbook marketing produced by government agencies or industry projected similar messaging on the role women played in improving family health. Thus, women were exposed to multiple mediums all projecting similar ideals, both empowering and restrictive, of their roles within the home.

Historians have long noted that advertising was a primary medium through which the discourse on women's role and power within the home played out, and few industries marketed confined gender norms more aggressively than the food industry.[5] Katherine J. Parkin notes in her 2006 book, *Food Is Love*, that despite the radical changes during the twentieth century in women's societal roles at home and in public, food advertisers almost exclusively marketed their products to women while continuing to project a static depiction of women as wives and mothers engaged in traditional homemaking activities: "In spite of the countless changes that shaped food, the advertising industries, and American society generally, food advertisers demonstrated a remarkable continuity in their approaches."[6] Parkin sees food advertisers in the United States as utilizing several key messages, many of which are clearly evident in Canadian seafood advertisements during the first half of the twentieth century. Chief among them were, first, that good women expressed their love for their family through the food they bought, prepared, and served; second, women were entitled to free time and so needed to turn to convenience foods to lighten their burden at home; third, food shopping and preparation could solidify the family's place within the middle class; and fourth, wives and mothers were responsible for family health and well-being.[7]

Magazine and newspaper advertising in Canada followed similar lines to what Parkin has found for US-based marketing. Valerie J.

Korinek's assessment of the historiography and methodology of print
media studies, and particularly that of women's magazines, is useful
to this study because much food advertising during the interwar period
appeared in women's magazines.[8] In her history of the feminist message embodied in *Chatelaine* magazine during the 1950s and '60s,
Korinek rejects what she identifies as the historical myth regarding
women's magazines; she shows that *Chatelaine* fundamentally
changed during that period from "a thin magazine devoted primarily
to fiction, departmental features, and cheerful editorials" into an
important source of Canadian feminism. Yet too many scholars, she
maintains, uncritically dismiss women's magazines as patriarchal rags
produced by publishers who actively sought to manufacture insecuri-
ties among female readers to encourage them to buy advertised products to fill voids – a point of view that began with Betty Friedan and
has often been adopted since. Such a view, Korinek argues, "entirely
negates women's agency" and presents them as empty vessels to be
filled with whatever discourse or rhetoric an editor or writer wishes.[9]

Korinek suggests that historians of print media should be wary of
reductionist assessments of women's magazines and their readership.
Works like Helen Damon-Moore's *Magazines for the Millions*, Jennifer
Scanlon's *Inarticulate Longings*, and Gruber Garvey's *Adman in the Par-
lor*, she notes, all presume that, rather than participatory media, "the
magazines were 'prescriptive' products that taught middle class women
how to behave." She warns about "overly deterministic analysis of the
impact of magazine publishers," and asserts that historians need to better assess readers' reaction to the stories and advertisements presented
in these magazines. By writing letters to the editors, she says, women
readers became part of the magazine's community, active contributors
in shaping its message. In demonstrating that writers and editors
eagerly sought and incorporated feedback into their production of
content, she characterizes *Chatelaine* as "participatory."[10]

But it is exceedingly difficult to find the two-way communication
between publisher/editor/writer and reader within advertising decisions and messaging in magazines and newspapers that Korinek finds
within the *Chatelaine*'s editorials, articles, and features.[11] Even in her
own assessment of advertisements in *Chatelaine* during its feminist
era of the 1950s and '60s, Korinek notes that they remained one-
dimensional and "uniformly depicted a comfortable world of middle-
class consumption ... Whatever the messages of other editorial
sections of the magazine, the advertisements continued to depict

women's roles very narrowly."[12] The codes that Korinek identifies in
the ads – superwoman, romance and marital bliss, suburbia, femininity, dieting, youth, sex – all depict a "middle-class suburban utopia"
contradictory to the feminist messaging she found in the editorials
and stories. Throughout the twentieth century, food messaging in
print media remained conservative, and Korinek's assertion of participatory discourse between publisher and reader does not seem to
apply to advertisements. She concludes her section on advertising by
providing analytical models that show readers did not often read ads
closely, nor did they scrutinize advertising messages to the same
depth as those in stories.

Perhaps advertising marketing seafood in Canada was not as successful as advertisers would have hoped because they failed to incorporate a more participatory model. It is difficult, nearly impossible,
to get a sense of how readers responded to the seafood advertisements the Department of Fisheries and the Canadian Fisheries Association placed in the nation's magazines and newspapers during the
interwar period. The archives contain only a few responses to advertisement messaging, and so any assessment of consumers' reaction to
it is necessarily limited to changes in sales of the product. That measurement is, of course, flawed, because consumers make their purchasing decisions based on factors other than advertising. Seafood
advertisers generally projected an image of the happy housewife
making informed decisions dedicated to the well-being of her family. If, as the evidence suggests, the advertising of seafood failed to
stimulate increased purchases and consumption, then perhaps we
can conclude that women rejected this advertising message. Yet
we are still at a loss as to why. Did they object to the imaging and
messaging? Did they disagree that they needed to add more seafood
to their family's diet? Or did they just not like fish? In the end, the
evidence is vague and circumstantial, and we are left with assessing
only the message and not how it was received.

The messaging about seafood from government and industry cookbooks and food advertisements specifically targeted the Canadian
middle class. The female shoppers that the Canadian Department of
Fisheries and the nation's fishing industry addressed during the first
half of the twentieth century lived in a period of radical transformation of the typical image of the home, increasingly seen as middle
class. Before industrialization, domestic work was central to the economic well-being and functionality of the home and family. After

industrialization, the home took on more moral than economic value. As middle-class women fell out of the field of income-generating "work," they were increasingly tasked with making the home perfect – more divine. By 1900, this "domestic theology" was well entrenched in Canadian culture.[13] Food shopping and preparation were central to this divinity as middle-class women assumed the predominant role in defining the family diet and that diet came to be seen as important to both physical and moral health.

Like historians in the United States and elsewhere, Canadian historians have had difficulty in defining the middle class and instead often focus on middle-class values, ideologies, or movements. In his 2000 history of the Ontario towns of Galt and Goderich, Andrew Holman notes that, in general, "historians have employed the term [middle class] more as an adjective than a noun"; yet as his study illustrates, a body of people "of the middling ranks came to see themselves as a distinctive social structural unit rooted in local socioeconomic realities." Like many other historians of class, Holman focuses primarily on economics – occupation and income – in identifying a middle class, and, further, emphasizes the importance of industrialization in its formation. Between the 1850s and 1890s, he contends, Canada began transforming from a rural agrarian society rooted in Old World social ordering into a modern industrial nation-state, which provided the context for "an identifiable and self-identified middle class in Canada."[14] For Holman, Canada's middle class was more than just an ideology – it was a demographic reality: "The middle class in these places [Galt and Goderich] was composed of real, functioning, and integrated groups of people for whom a collective identity and an identifiable set of values had considerable local meaning and use."[15] Middle-class values encompassed a host of ideologies or behaviours that included engaging in social reform, moralism, religion, gender norms, education, and child care – to name a few – but for the most part, middle-class status came down to occupation and income.

"Occupation was the principal determinant of class status ... In short, the middle class included those who worked for a living, but not with their hands, and those who did work with their hands, but also owned their means of production, perhaps employing others."[16] Businessmen, professionals, and white-collar workers made up the Ontario middle class so central to seafood marketers of the early twentieth century. Holman does recognize that a definition limited by occupation alone leaves little place in the historiography for middle-class women, for to be in the middle class almost always meant that the sole breadwinner was the patriarch. As such, wives, daughters, and sisters only achieved middle-class status via their husbands, fathers, and brothers. Holman notes, however, that women could construct their own class identification through various associations and organizations, most often those dedicated to shaping the moral order of society.

Certainly, occupation and income were key elements in defining class, yet cultural behaviours mattered as much as earning power. In his 1990 study on the middle class in English-Canadian universities, Paul Axelrod discussed the theory of class identification by categorizing the scholarship into two dominant veins. Marxist theory defines class by narrow objective identifiers, namely, the ownership of the "means of production." "Social stratification" theory, however, uses both objective and subjective criteria such as income, education, standard of living, and the ill-defined concept of prestige to define class differentials. Neo-Marxism, or socialist humanism, beginning with Edward P. Thompson, overcame Marx's rigidness by recognizing that class is not static but a phenomenon shaped by collective identity and action. Still, both Marxist and neo-Marxist theorists on class are often too trapped by occupation and income statistics and fail to recognize cultural characteristics in class identity. It was these cultural tropes – family, health, civic virtue, and so on – that seafood marketing mostly focused on; advertisements made virtually no references to consumers' occupations. These more subjective analytical points used by the stratification theorists have not gone without critique. As Axelrod notes, "Stratification theorists are sometimes accused of being purveyors of mythology and apologists for the status quo because their work implies that individuals, whatever their social origins, can attain upward social mobility by dint of proper attitudes and diligent work." While the truth of such abilities to improve social position via attitude and work may be up for some debate, it is nonetheless true that many Canadian shoppers in the interwar years believed that such mobility was possible via consumption behaviour.[17]

For Axelrod, the key elements in class definition went beyond occupation and included "economic position, collective class consciousness, and perceived social status."[18] Important objective indicators, such as economic wealth, are obviously important to understanding class, yet as Axelrod notes, "once one's 'objective' class position has been determined, however, one's values and self-image deserve careful attention if the complex dynamics within the social class are to be understood."[19] In the end, Axelrod defines middle-class families "as those whose major income-earners were non-manual workers who enjoyed social status but exercised limited economic power, and whose standard of living ranged from the very modest to the very comfortable."[20] The period of industrial growth at the turn of the twentieth century was, for Axelrod, the historical era most important to the proliferation of a Canadian middle class. He argues that the new industrial order that emerged then created a new social and economic class structure; "A variety of newer professions augmented the traditional ones, reflecting active interest in applied science, growing concern about public health, and widespread efforts to regulate both civic affairs and the behaviour of the working class, immigrants, and the poor," setting the context for "a 'new' middle class."[21] These were the shoppers that most producers most actively sought out via advertising campaigns. Since university education was often essential for these new middle-class occupations, Axelrod concludes that it was this middle class that came to dominate university attendance and it was university attendance that most defined middle-class identity. Yet consumer behaviour and gender norms certainly were equally important as university education and professional occupation as shapers of middle-class ideology.

While Holman sees a Canadian middle class in 1850s–60s Ontario, and Axelrod finds them in the turn-of-the-century English-Canadian university, Franca Iacovetta uncovers how central Canadian middle-class ideas were in shaping post–Second World War refugee and immigration policy. Although chronologically outside the scope of this study, Iacovetta's definitions of "Canadian ways" are helpful in illustrating Canadian cultural values in the interwar period. The country's dominant cultural norms, she maintains, were influenced by "everything from food customs, child rearing, marriage, and family dynamics to participatory democracy, kitchen consumerism, and anti-Communist activism" and were all grounded in middle-class ideologies.[22] These ideologies were certainly not new in the late 1940s and early '50s but were the end result of decades of economic transition to industrialization and the cultural revolution known as modernism. The "dominant paradigm" that Iacovetta identifies as the source of middle-class "gatekeepers" in 1940s and '50s Canada was equally powerful in 1930s and '40s consumerism. She notes that this middle-class hegemony "normalized conventional bourgeois ideals of proper gender roles, the family, and sexual behavior and sought to cultivate good citizens who would be as cognizant of their civic duties to the state and wider society as of their individual rights and social benefits."[23] The bourgeois ideals that Iacovetta lists, especially those related to gender, family, and civic duty, all appear in seafood advertising that emphasized women's domestic roles, the central importance of a nuclear family unit, and consumers' responsibility to buy national product and support Canadian industries. Consumerism had emerged as not just a key economic structure but also a key middle-class cultural ideal.

The early to mid-twentieth century not only saw the industrialization of Canada and the redefinition of household work but also witnessed the transition to a consumer-based economy – the third of the four major themes at the intersection of food and women's history relevant to this study. Consumption, not production, defined national economic well-being, and in this vein, women assumed more cultural power as primary shoppers for the family. Industry success increasingly relied on a company's ability to convince the nation's women to buy its products. In the world of consumer economics, women were now acquiring their consumer knowledge from a flood of advertisements, magazines, cookbooks, housekeeping texts, and mother's manuals. By 1900, Canadian women faced a revolutionary modern age so fundamentally different from their parents' world that it appeared to many young women setting up a new household that their mothers had little good advice to offer and so increasingly turned to other sources of knowledge. Thus a whole new way of looking at housework, including food shopping and preparation, opened new opportunities for food producers. Within this context, seafood might gain entry even into homes distant from the coast if these new methods of knowledge transfer could convince the modern middle-class wife and mother that seafood was part of the new wholesome and modern diet.

The food knowledge of the Canadian middle-class woman, as well as her consumption habits, increasingly came under the scrutiny of Progressive-era ideals of logic, science, and efficiency. The Progressive influence over the home had its most profound manifestation in the discipline of home economics. Laura Shapiro in her 1986 book, *Perfection Salad*, outlines the influence of the domestic-science movement and, after 1899, home economics, on shaping how people understood food. Both domestic science and home economics, she notes, fit within the context of Progressive reform, a diverse movement of middle-class reformers determined to fix the ills of society via the professional and pragmatic application of scientifically derived solutions. That trend towards professionalization, with the communication of information on calories and vitamins, fundamentally altered American – and Canadian – food consumption behaviour.

Although concerned with a wide variety of issues facing women in the homes, female advocates of home economics focused overwhelmingly on the pragmatic use of food. They concerned themselves with the preparation of nutritious, convenient, and inexpensive meals in which taste and cultural habit mattered little, if at all.[24] The home economists of the early twentieth century rejected tradition and assumed that the ways of their grandmothers were outdated, if not barbaric. This transition in food shopping, preparation, and consumption redirected authority outside of the traditional, generationally defined home in which daughter learned from mother. Instead, young, mainly middle-class, aspiring mothers had to turn to other sources of authority. Government agencies, universities, and industry all exploited this breach in knowledge transfer and inserted themselves as the primary sources of food knowledge. After 1900, women increasingly turned to the "divine" authority of pragmatic science, as researched, tested, and communicated by the three branches of that divinity: government, university, and industry. Much of this knowledge was transferred through advertising and government- and industry-funded cookbooks and cooking tours.[25] By the end of the First World War, there was a fundamental reorientation of how women acquired food knowledge and how they used it to shape the middle-class family's diet via consumption of modern foods.

Shapiro notes the important role that women themselves played in this transition. A relatively small group of home economists gained significant power in the culture of food and built strong alliances with government, university, and industry. The doors of power opened to women, but did so very narrowly, allowing them to play only marginal roles within preconceived notions of their proper place. Though the domestic sphere was extended, it was still largely defined as the domestic sphere. Women could become food nutritionists but not chemists. This fit well within the Progressive era, as these women were one small faction of a broader movement to deal with the generally accepted ills of modern industrialization. Shapiro notes, "By [female home economists'] reckoning, domestic science was a more radical solution than socialism to the problems of urban poverty, and a more visionary response than feminism to the indignities suffered by women."[26] A better-fed family meant social stability, and a more informed mother was a more powerful woman. Yet as Shapiro argues, history must conclude that home economists largely failed in their effort; they tried to break gender roles while working within those limited structures. According to Shapiro, few home economists gained any authority within government, university, or industry before the 1950s, and they did not seem to inspire any blossoming of female scientific research. Most home economists remained chiefly communicators, distributing information derived from male-led scientific research to a predominantly female-led consuming public. Furthermore, the culinary ideal that they championed was overwhelmingly bland and, as Shapiro argues, is the root of much of the criticism lodged against the American cuisine of the mid-twentieth century, dominated as it was by conveniently packaged and largely flavourless food products void of any local or culturally defined value.

Shapiro's assessment fits well with the history of seafood marketing in Canada. Although many women played important roles in communicating the government and industry message, few participated in the shaping of that message. Neither did women achieve long-term employment within fisheries bureaucracy until the 1950s, and only then in limited roles as professional home economists.[27]

In contrast to the importance of women in dictating seafood consumption, the iconographic image of North American fisheries production has long been the masculine, independent, and daring fisherman. This imagery, captured powerfully by artists and writers like Winslow Homer and Rudyard Kipling, dominated the public perception of the fisheries in both Canada and the United States throughout the twentieth century.[28] Yet, beginning around 1910, the Department of Marine and Fisheries in Canada, along with governmental allies like the Canada Food Board during the First World War, and industry allies like the Canadian Fisheries Association, began to fundamentally recast the iconographic images society associated with the fisheries. Alongside the slicker-clad, often grisly fisherman appeared the dainty housewife providing meals for her loving family.[29] Neither image was accurate, but accuracy, or reality, has little to do with consumer culture. Advertising cast mothers as dutifully creating wholesome, economic meals, enshrining their families within the middle class via their purchases, and rescuing the fisheries with their power of consumption. As the fisheries economy continued to decline throughout the twentieth century, with the exceptions of wartime booms in 1916–18 and again in 1939–45, fisheries bureaucrats and industry leaders increasingly turned to Canada's housewives not just as a potential solution to the decline, but, surprisingly, as a cause of that decline in the first place. In the advertising documentation and internal communications within and between government and industry leaders from 1900 to 1940, housewives were blamed for the failure of the nation's fisheries. In the opinion of government and industry leaders, Canadian women simply did not know how to properly select, prepare, and serve seafood; therefore, through ignorance or laziness, they polluted public opinion about the value and tastiness of fish as food and ruined an economy of great national importance. However, via a large-scale, well-organized, and sustained "educational" campaign, these women could be taught the value of fish as a food and trained to incorporate it into a balanced family diet (figure 3.2). This would dramatically increase national consumption of seafood and save the industry from its prolonged and at times seemingly inescapable decline.

Advertising of consumer goods more generally targeted women as the primary purchasers.[30] Likewise, within the food industry, producers understood that the nation's housewives were the primary shoppers for food products and largely defined what families ate. Seafood producers used this assumption to convince grocers to make direct appeals to women. The pages of the Canadian Grocer are filled with advertisements touting their product's appeal to the female customer. Many ads noted the housewife's ability to identify food quality and her celebrated particularity about what she fed her family. Processed seafood producers like Connors Brothers and Thistle Brand called upon grocers to make personal connections with their female customers and guide them in making wise food selections. Connors Brothers noted in a 1915 ad that "a stock of Brunswick Brand Sea Foods will enable you to satisfy the most particular housewife in your town."[31] Thistle Brand too insisted housewives knew quality, and grocers who catered to that demand would benefit: "Years of satisfactory use by the Canadian housewife have educated her to look for the Thistle Brand when buying fish canned or cured."[32] Seafood producers still relied on grocers and jobbers to push their product rather than engaging themselves in the direct-

LE POISSON se prête à la préparation de mets tentants

et délicieux—dîners merveilleux—soupers satisfaisants . . . goûters appétissants. Vous pouvez facilement préparer une grande variété de menus puisqu'il y a plus de soixante différentes sortes comestibles de poissons, de mollusques et de crustacés canadiens—dont la plupart sont disponibles toute l'année durant.

Régalez votre famille avec ces merveilleux plats de POISSON dont la préparation est si facile. Ecrivez aujourd'hui pour avoir votre brochure de recettes. Servez souvent du poisson—et les membres de votre famille vanteront vos talents de cordon bleu.

MINISTÈRE DES PÊCHERIES, OTTAWA.

MINISTÈRE DES PÊCHERIES, OTTAWA.   924
    Veuillez m'envoyer votre brochure de 57 pages intitulée "100 Délicieuses Recettes de Poisson".
Nom..
            (Veuillez écrire lisiblement)
Adresse..
                                             RP-IIF

LE POISSON AU MENU DU JOUR

La vente suit l'annonce, elle ne la précède pas

Fig. 3.2   "Les Repas de Poisson," *L'Action Paroissiale*, February 1940, 19. The Department of Fisheries' French- and English-language ad copy carried identical messaging, emphasizing the health qualities of fish via the established gender ideals of the period, charging Canadian women with maintaining the nation's public health.

consumer marketing already common in other food sectors. Grocers had to be convinced that selling to women would be profitable for them by establishing a regular base of trade. "The housewife's appreciation of Thistle Brand is evidenced in the heavy sales you will experience the year round,"[33] insisted the Montreal-based wholesaler, while Connors Brothers pointed to the "heavy and continuous demand" created by "the enthusiasm of the housewife and the regularity with which she comes back for more."[34]

Seafood advertisements worked within established tropes of housewife consumerism. Ability to identify quality was a main one, as was the pressure women felt to provide a variety of meals for demanding families. Amidst an onslaught of food advertising during the 1910s and 1920s, more and more people sought diversity and turned to the wife and mother to provide it. Women had to navigate a growingly complex food market to meet growingly complex family food demands. Diversity of food was particularly important to an emerging middle-class identity, and advertisers bought into this anxiety, and even propelled it, endorsing housewives' responsibility to avoid monotonous family diets that could prove nutritionally and emotionally detrimental. Fish, many seafood companies insisted, could be a core item in the goal for varied meals. Hamilton-based food wholesaler John W. Bickle and Greening noted, "It is hard for her to decide what to get for some of the 1,000 meals a year that the average housewife has to plan." The company placed the burden on the grocer to offer educated assistance: "She depends upon you, Mr. Grocer, to suggest something, why don't you tell her about King Oscar sardines?"[35] Connors Brothers, too, asked grocers to provide direct, informed suggestions to women struggling to diversify menus. In the face of a wide variety of breakfast cereals, the company advised a return to a traditional breakfast of sardines: "Yes, she'll surely appreciate your suggestion if you make it Brunswick Brand. Most housewives know the appetizing goodness of Brunswick Brand Sea Foods for luncheon or dinner. But for breakfast – well, now that IS a capital suggestion."[36]

Government marketing campaigns during the 1930s continued to highlight the value of variety and fish as the solution to women's conundrum. In a 1936 advertisement, the federal government claimed, "There is not a woman in Canada today who has not occasionally found herself at the end of her tether ... wondering what on earth to serve for the next meal. Yet if she thought more often of Fish, her problems would be over."[37] A radio commercial carried throughout Atlantic Canada highlighted the "teeming variety" of fish available.[38] Thus, both the government and private industry bought into established idealizations of housewives as informed consumers eager to provide for demanding families in a changing modern world.

In the new celebration of home economics, women became informed not just about food quality but more complex issues of budgeting. Canned fish had a long-established association as inexpensive, and in times of economic hardship, advertisers like Connors Brothers marketed their product as cost-saving and targeted another idealized characteristic of the housewife, that of thriftiness. A 1916 Connors advertisement directly compared sardines with meat: "The High Price of Meat is Turning the housewife to Sea Foods and the alert Grocer is utilizing the situation to his profit by featuring Brunswick Brand Sea Foods."[39] During the inflationary year of 1917, the company again urged grocers to help the nation's women make economically informed decisions: "The increasing cost of foodstuff is receiving marked attention from the thrifty Canadian housewife. Her mind is bent on evolving a menu for her household that will give the greatest food value at the lowest cost." In the same ad it pitched its leading brand: "Brunswick Brand is surely the ideal food for the thrifty housewife, and one whose sterling qualities she will quickly recognize."[40] During the war years, thriftiness was celebrated not just for family economic livelihood but also national security. A 1917 Connors Brothers ad proclaimed, "Housewives everywhere who are doing their bit to conserve the nation's food supplies are fully alive to the great economy of using good, wholesome sea foods extensively."[41]

Yet advertisers had to be careful that marketing the economic value of canned fish did not suggest sardines were poor-quality food or that seafood more generally was associated with a low-income diet. Throughout the 1910s and 1920s, they sought to elevate the sardine from the worker's lunchbox to the middle-class table by celebrating its inexpensiveness in a way that did not suggest it was the food of the poor. Connors Brother took pains to separate low price from low quality by emphasizing economic efficiency: "More and more careful and efficient housewives are learning that Brunswick lines stand for *Superior* Quality."[42] Their ads elevated the idealized thriftiness of housewives to economic intelligence: "Brunswick Brand Sardines are easier to sell because their price appeals to the economical housewife, who must make her dollars go as far as possible."[43] She purchased sardines because she was an intelligent consumer, not a poverty-stricken one desperate for cheap food. The repeatedly used phrase "economical housewife" recast her as an informed consumer calculating her budget to secure the family a footing as middle-class Canadians.[44] The company's advertising educated the Canadian housewife on broader concepts of economy and emphasized that sardines' inexpensiveness was unrelated to lower quality. And unlike imports, Canada's sardines came to the consumer free of excess taxation and logistical expenses; thus, an advertisement claimed, "We are enabled to save money for the Canadian housewife, increase her purchasing power for other lines."[45] Women who bought sardines could redirect family funds to other consumer ventures, more easily placing their families within the emerging middle class.

During the Depression, the idealization of home economics took on added meaning for housewives responsible for family survival through the nation's harshest economic downturn. Government sponsorship of seafood advertising in the mid-1930s highlighted women's ability to incorporate inexpensive, easily prepared meals intelligently into shrinking family budgets: "Fish, unlike meat, may be bought at cheap prices and still be of the first quality. Housewives with limited budgets who are pondering the needs of alert young bodies, here is good news for you!"[46] Housewives were called upon to provide health and economic security for their family and the nation. By wisely buying inexpensive seafood, they were told, they could reduce food costs and avoid the appearance of having fallen into poverty and needing to purchase low-grade meat.

It is difficult to determine the extent to which consumers accepted the recasting of seafood, particularly processed canned seafood, as economical rather than a food of the poor. This effort was somewhat undermined by the reality that many seafood products were comparatively expensive. Some Canadian would-be consumers took the time to write to the Department of Fisheries disputing its claims that fish was inexpensive food. Arthur E. Chant of Regina, for example, wrote in February 1939 that consumers didn't need new recipes for fish: the real "trouble is cost." Chant enclosed the "Any Day a Fish Day" ad clipped from his newspaper with the words "This is too much," and "Good Lord" over recipes for Stuffed Fish a La Newburg and Creamed Fish in Hot Biscuits.[47] Mrs Dana F. Lucy of Calgary declared in a July 1940 letter that though she belonged "to the very average wage earning class," the department's fish recipes were simply too expensive for her to prepare. She wanted to buy Canadian goods to help Canadian industry but could not find the "'new low prices'" referred to in the government ad.[48] Similar letters complained that price made seafood "prohibitive for the majority of families," and that "higher prices of course will place this in a luxury class for the average person."[49] The *Halifax Herald* echoed these consumer responses, questioning the government's campaign in 1937: "The rising cost[s] of living are worrying many housewives and they are looking for the very best value possible for their money ... [but] ... How can the housewife buy when she has to pay such prices?"[50]

By controlling the cost of food purchases, women played important roles in family survival during economic downturns. In food preparation as well, advertisers recognized the importance of catering to gendered assumptions about women's role. Ease of preparation became a key selling feature. The time saved meant partial liberation from the kitchen. Connors Brothers insisted Brunswick Sardines were "clean, cooked, and ready to serve," without "the muss and fuss of cleaning and preparing"; there was "nothing to do but eat."[51] They "solve[d] for the housewife the difficulty of preparation and serving, and ensure[d] the very best quality that can be had."[52] Marketing ease of use was particularly popular in the summer months when no one could "expect a woman to enjoy being stuck in a kitchen." Sardines were "the very thing to give her nourishing meals without the bother and fuss of tedious preparation," allowing more time to enjoy the "cool out-of-doors."[53] Women looking for "food stuffs which are delicious and wholesome, yet require no standing over hot stoves in their preparation," found it with Connors Brothers' canned fish.[54]

Underlying the marketing strategies around ease of use was the early twentieth-century emphasis on a more active and youthful culture. As "outings" were increasingly important for young couples, women sought more convenient menus for on-the-go family activities. *Canadian Grocer* noted this trend in a 1926 editorial: "More and more the general public is commencing to look upon the summer outing as a needed stimulant." Although the new availability of the automobile was a factor, it was part of the general trend to "take better care of our bodies that we may be more fit to meet the responsibilities of life." Healthy foods for picnicking, camping, driving, or just something quick before an outing took on added value: "There are many occasions in every household when a tasty meal is wanted, but no time to prepare and cook it."[55] "Summer days are sardine days," another Connors Brothers advertisement declared – the perfect choice for combining healthy activities and healthy eating.[56]

Emphasis on ease of preparation carried into the 1930s. Industry cooperatives and the government continued to sell fish as a simple meal. In 1931 the Canned Salmon Advertising Committee hired Florence Robertson to deliver a series of radio shows highlighting the convenience of canned salmon that "relieves the housewife of all the troublesome preparation necessary with freshly caught fish."[57] During the government-funded campaigns of the late 1930s, H.F.S. Paisley, director of Fisheries Intelligence and Publicity within the Department of Fisheries, advised the advertising firm E.W. Reynolds that the recipes they had developed were too complex: "In general we want the suggestions to housewives to cover simple dishes easily prepared and with a reasonable minimum of accompaniments which would add to the cost."[58]

The general celebration of the housewife as informed consumer contrasted sharply with the concern among fishery industry leaders and government bureaucrats that women were poorly informed fish buyers. Among these men, the idea that women simply knew nothing about fish became increasingly dominant in the 1920s. In 1928, Alfred H. Brittain, chairman of the Publicity Committee of the Fisheries Association, called for government-funded advertising that specifically addressed the lack of fish knowledge among Canada's housewives: "What, for example, does the average housewife in an Ontario town know about the best method of preparing the various varieties of sea food for the table? ... With real fish stores to display fish, and knowledge on the part of cooks to prepare them in the most attractive way for the table, there would be a greatly enlarged demand for this class of food."[59]

The industry viewed the inability of women to properly select, cook, and serve fish to their family as a significant problem. By the 1920s, many industry leaders believed that if they could better market fish to the nation's housewives, they could save the fisheries from their economic woes. So B.K. Snow, another chairman of the Publicity Committee of the Canadian Fisheries Association, stated in a 1926 letter to an industry partner, "We feel that if the demand is created from the housewife and the consumer that the distribution will necessarily follow."[60] By the 1930s government bureaucrats too saw the nation's women as both cause and solution for the depressed state of the country's fisheries. As Deputy Minister of

Fisheries William Found wrote in a 1936 letter, "After all it is in the main the housewives of the country who must be influenced if we are to increase the consumption of Canadian fish foods within the dominion."[61]

The emerging belief within industry and government was that some blame for fisheries' poor economic productivity could be laid at the iconic housewife's feet. She had failed to properly utilize the nation's valuable natural resources. In times of war, housewives' lack of engagement with seafood took on increased national security importance on the basis that low domestic consumption could actually undermine war efforts. During the First World War, the Canada Food Board (see chapter 1) initiated widespread marketing of Canada's fish to relieve pressure on meat and meat products. Some of this marketing directly pointed to the Canadian housewife as a cause of under-consumption of seafood. During the National Fish Day campaign of 1918, for example, the Canada Food Board released an article calling upon Canadians to "Eat More Codfish" that partially faulted women: "Because Canadian housewives have proved indifferent to their food value, the bulk of this amount has been salted and dried in the past and sent to the Latin countries of Europe and to South America."[62] The observation suggested that Canadian fisheries were forced to find export markets – notoriously unreliable during wartime – because Canadian women did not appreciate the value of fish as a food.

The perception of housewives as a core problem continued throughout the early twentieth century. In a 1931 letter to William Foran, secretary of the Civil Service Commission radio series, William Found was direct in assigning blame: "One of the main reasons, indeed probably the only one, why fish are not more commonly used is because the best methods of cooking and preparing it for the table are not generally known by the housewives."[63] The following year, Found appeared on the commission's radio series, intended to introduce Canadians to the diversity of civil service work in the federal government, and noted the importance of educating women on the properties of fish and proper use of seafood. After a thorough review of the Department of Fisheries and its work to aid the industry, he closed his talk: "Being satisfied that increased use of fish as food would speedily result if our housewives were more generally familiar with simple, economical and at the same time highly palatable methods of cooking fish, the Department is undertaking what may be described as an educational campaign with this in view."[64]

Radio became a key medium to reach the nation's housewives directly in their kitchens. In October, Found wrote to J.G. McMurtrie, broadcasting manager of the Canadian National Railway's radio department, asking if there was a way to make more references to fish as food during their Tuesday and Thursday home show. "Our object of course is to aid in the further development of an important Canadian natural resource, but we would think it likely that a brief talk on fish cookery from time to time would be of interest to housewives who listen to your broadcast."[65] Again, development of this natural resource depended on housewives' cooking ability. McMutrie was happy to comply, noting the project's national importance. If CNR broadcasts could aid in the increased consumption of Canada's fish, it would "also automatically increase the Country's business which includes our own."[66] Industry growth was the objective, and improved consumer knowledge, especially among women, was the means.

Found was also able to provide promotional material to Iola Plaxton of the Radio Specialists of Canada in Toronto. The "educational fish propaganda," Plaxton wrote, would be used to inform Canadian housewives of the value of adding fish to family diets. The radio talks would "do much to increase the daily consumption of fish, as we are giving valuable information in connection with the home."[67] Found included in the material the brochure "Food and Health Value of Fish," prepared by the Department of Fisheries for circulation throughout public media. The text focused on the healthfulness of fish as a food, the economic value of adding more fish to the family diet, and the wide range of seafood available to vary breakfast, lunch, and dinner menus.

Industry leaders too turned to radio broadcasts to educate Canadian housewives. Associated Salmon Packers of British Columbia sponsored *Symphony of the Sea*, a program that aired in 1929 on Thursdays from 10:00 a.m. to 11:00 and was "particularly for the housewife and designed to acquaint her with the many desirable qualities of pink salmon and its place in a well-balanced diet."[68] The BC salmon industry faced economic challenges in the 1910s and '20s, especially in lower-grade lines like canned pink salmon. After repeated failures to secure a better footing in foreign markets, it increasingly turned to a focus on domestic consumption and gender-based marketing.

By March 1930, the department was ready to initiate its first organized and sustained radio campaign targeting women. It hired Mildred Campbell to deliver a series of talks on the Tuesday-morning home shows on the CNR radio broadcasts. Campbell (as noted in the previous chapter) was a talented fisheries scientist working on her doctorate, yet the department valued her largely as someone who could communicate to Canada's housewives the importance of fish as food. Although previous campaigns were openly designed to help the industry, this series claimed to have an altruistic ideal of providing a public service. In her first address, Campbell explained, "They will simply be informal little talks about fish as a food, why to cook it and how to cook it, with a view of giving our Canadian housekeepers additional information as to fish cookery and suggested recipes for serving fish in tasty attractive ways."[69] In a subsequent broadcast, she said her talks were "simply in an effort to be of service to Canadian housewives."[70] Occasionally she touched on nationalist issues, such as the superiority of Canadian fish over foreign imports, and, more rarely, the history or contemporary technology of the fisheries. However, the content was predominantly the nutritious value of fish and recipes for fish dishes.

Campbell touched on many of the themes identified above, primarily the health qualities of seafood, but she also addressed the range of product available to the busy housewife providing family meals in at a time of economic uncertainty: "There is no lack of variety for the Canadian housewife who wants canned fish."[71] Moreover, "the diversity in the ways in which it may be cooked makes it of exceptional helpfulness to the housewife in avoiding that bane of the table – monotony of diet."[72] Thus, Campbell's listeners could support middle-class eating habits (variety) while also assisting national economic growth by using more of Canada's own food resources.

For some government and industry leaders, the problem was not just a general lack of knowledge among Canada's housewives about fish but their inability to cook it successfully. In 1924, Thomas Blanchard, of the Blanchard Fish Market in London, Ontario, expostulated, "It is time housewives were educated to keep the blame frying pan in its place."[73] Improving housewives' fish cookery became the singular focus of Evelene Spencer, hired by the Department of Fisheries in 1932 to embark on a nationwide tour. William Found, who hired Spencer and directed her activities, stressed the importance of women's cooking to national economic recovery; the department was "satisfied that increased use of fish as food would speedily result if our housewives were more generally familiar with simple, economical and at the same time highly palatable methods of cooking fish." Found publicized Spencer's tour in his radio addresses and "hoped that our housewives and those anticipating that they will occupy that highly honourable and outstanding position, will not fail to hear and heed her."[74] The idealization of the housewife in his statement, however, contradicted the message of misinformed women in need of the help of the gentlemen of the Department of Fisheries.

Spencer herself believed adamantly in educating women on proper cooking in order to stimulate economic recovery in the fisheries. She wrote as much to Found in 1931: "The ultimate goal of this campaign is to increase the consumption of fish. The education of the housewife is a strong factor in reaching this."[75] The Department of Fisheries paid her $175 a month and covered her expenses along with those of her assistant, her daughter; it also provided the services of D.H. Sutherland to make pre-arrival arrangements for lodging, demonstration events, and meetings with local department stores, women's groups, journalists, and industry representatives. The fisheries industry, through the Canadian Fisheries Association, agreed to provide Spencer with as much fish as she needed for her demonstrations. Local representatives of the department and the industry worked to create as much buzz as possible before Spencer's arrival in town.[76] They covered stores with banners and streamers, contacted newspaper reporters, and met with community groups and department stores to arrange special events and sales to coincide with Spencer's talks.[77] Their efforts succeeded in generating large crowds to listen to her presentations and her demonstrations on her specific high-heat baking method, widely known by the end of the tour as the "Spencer Method."[78]

Spencer had long been preaching better cooking methods. In the 1920s she had worked to help market Nova Scotia seafood products in Quebec and Ontario, particularly fresh and frozen fish fillets caught by the province's young trawling fleet. In a 1925 interview with the *Toronto Star*, she explained, "What I am trying to do is teach the women the use of filleted fish." She celebrated the idea that the fillet eliminated the need to clean the product before cooking: "Fillets have now put fish in the beefsteak class – there is no waste, no bones, no fat, just straight meat."[79] During her tour, however, the Department of Fisheries directed her to teach women the value of *all* the nation's fish products without bias towards any one industry sector, although she could specifically highlight whatever fish product was most conveniently available in the area where she was working.

Spencer's national campaign reflects one of the key shortcomings of the heavy reliance on government-funded marketing. To avoid accusations of bias through references to particular species or types of fish products, Spencer used vague, generalized references. Fish were homogenized; this limited her ability to educate consumers about the product, and perhaps forced her, and other government marketing efforts, to focus on creating new images of fish consumption and fish consumers rather than new understandings or appreciations of seafood. Spencer thus continued to rely on the ease of meal preparation for busy families, remarking in a Montreal *Gazette* interview, "Nothing, except an egg, is so quickly cooked as fish. This tip is for the late-returning housewife faced with the problem of having a meal ready for the head of the house when he arrives a few minutes later."[80]

Spencer provided Found with detailed reports on her activities. Her reports were informal in tone, often long, and more focused on her travel troubles than specifics of her work. They contrasted sharply with Found's replies, which were usually short and professional in tone, never addressing or replying to personal matters. From July 1932 to April 1933, Spencer was in Manitoba, Alberta, and British Columbia, presenting before crowds of several hundreds. In 1934, she was in Montreal giving demonstrations on mass feeding at hospitals and welfare kitchens. She next went to Toronto and Hamilton in southern Ontario.[81] Throughout, she worked closely with local women's groups, journalists, department and grocery-store operators, and industry partners, obtaining local supplies of seafood for her demonstrations to show audiences they could easily duplicate her efforts. (These local efforts, however, had to fit within more generalized national messaging.)

Her talks, both on radio and in person, appear to have been consistent in content. She focused on cooking method, calling the frying pan an "obsolete article in fish cookery."[82] Improved cooking methods addressed what she felt was the greatest handicap to increased seafood consumption, and thereby industrial recovery: the monotony of fish dishes. A 1934 market survey by ad agency Cockfield, Brown supported her observations. The company found that 66 per cent of housewives cooked salmon in only one way, 75 per cent said the same about codfish, and 77 per cent had only one recipe for haddock.[83] Although Spencer occasionally referred to the health-giving qualities of fish, her primary focus was on improved cooking method and crafting a more diversified family diet. When she did discuss health, she relied on general references to advances in health science: "Now there comes such a strong health appeal with the urge to *eat more fish* that if we are just and fair minded we cannot disregard this without at least listening to the arguments advanced by many of the leading research chemists on the continent."[84] She did not go into depth on these advances, relying instead on vague references to vitamin D and fish oils and fats.

Spencer became ill in the fall of 1934 and died while in service to the Department of Fisheries in January 1935. She would have continued her work until that spring by going to Atlantic Canada, and Found tried unsuccessfully to have her contracted salary paid to her husband. In general, Spencer's work had been well received by industry and government, which mourned the loss of the "Fish Evangelist" who had attracted large crowds of both women and men.[85]

Although Spencer's focus on improving consumer knowledge on handling and cooking fish fit within the larger educational propaganda of the government and the industry, she was also concerned about the quality of supply provided to the consumer. She must have regularly heard criticisms of poor quality and inconsistency. In an article for the Canadian Fisheries Association's trade journal, *Canadian Fisherman*, she noted her "unique position, being sort of a link between the housewives I am teaching on the one hand and the industry which supplies them, on the other. I can see both sides of the question." She asked the industry to do more to ensure high-quality fish, particularly to the retail market, the "last link in a long chain" that far too often did not properly handle, store, or display quality fish.[86] Perhaps, she hinted, the industry itself was partially to blame for the limited use of seafood by the nation's housewives. It is impossible to know what Spencer (who could be "one of the boys" when meeting with industry leaders) really thought of the male bias of the government and the industry, and the insistence that it was Canadian housewives who needed to mend their ways if the nation's fisheries were to recover. Her article at least suggests a more nuanced understanding of the problem, recognizing that poor quality of supply was a factor in incorporating more seafood into family diets. That insight, however, was not widely accepted by industry and government until well after the Second World War.

Unsurprisingly, government bureaucrats and the industry leaders were largely celebratory of their efforts to educate women and bolster the fisheries industry. The *Saint John Telegraph-Journal*, a pro-industry media outlet, noted the "new interest on the part of housewives in the family diet" generated by government and industry advertising, and congratulated Minister of Fisheries Joseph-Enoil Michaud, who "prompted this interest on the part of housewives" with a successful educational campaign.[87] The Second World War interrupted these campaigns, and it is difficult to assess their real success or failure. Although thousands of women wrote for free cook-booklets, the overall consumption of seafood in Canada remained relatively consistent, and it is doubtful that women embraced seafood more after the campaigns than before. Following the Second World War, however, the department seemed to recognize the importance of providing quality products for housewives to work with. It also expanded upon pre-war efforts to use women's intelligence and labour in marketing seafood and opened an internal Home Economics Branch.

By the outbreak of the Second World War, seafood marketers and their government allies had produced a series of images and rhetorical tropes that celebrated the nation's middle-class housewives. These images were unblemished by the challenging realities of a changing industrial world that ushered in an era of both modernism and antimodernism. The images of women and class buttressed a growing, if still fragile, national identity in Canada. Women were called upon to rescue one of the nation's most iconic and historic industries while securing their own family's place within the newly celebrated middle class. Tapping into gender and class stereotypes, seafood marketers pitched a product within clearly defined cultural boundaries.

Even as the Department of Fisheries and the fisheries industry combined to remarket fish as a modern, healthy consumer good competitive with other processed foods on grocery-store shelves, they simultaneously sought to recast the image of the seafood consumer. Liberating fish and seafood from its historical association with poor-quality, albeit low-cost, dried, salted, and pickled food relied on changing the image of not only the product but also the consumer. The recast image of the seafood-eater in Canada was dominated by modern rhetoric and imagery of middle-class women who increasingly bore the burden of defining family identity, which at least partially rested on what the family ate. The family diet had to meet the needs of individual and family health as it also promoted national health. Beyond this, women were called upon to rescue the failing industry, which argued that its troubles were rooted in women's inability to cook fish. Thus, as the marketing deployed an idealized and simplistic image of the modern Canadian consumer, it also burdened that consumer with industry recovery.

4

Eating Our Way Out of Depression: Stimulating Consumer Economics for Industry Recovery

Throughout the first half of the twentieth century, the Canadian Department of Fisheries invested in the advertising and publicizing of Canadian fish in both domestic and foreign markets, although these campaigns were most pronounced in the domestic market. At first sporadic and unorganized, they often reflected immediate and regionally specific needs. During the war years, for example, the Canada Food Board invested funds to encourage Canadian consumers to eat more fish to save meat for troops deployed in Europe. Following the First World War, the campaigns became slightly more consistent. In the winter of 1923–24, the Department of Fisheries marginally supported, with a grant of $10,000, the industry's corporate cooperative, the Canadian Fisheries Association (CFA), in its effort to get Canadians to "Eat More Fish." In the early 1930s, the department hired biologist Mildred Campbell to deliver a series of radio talks encouraging housewives to utilize fish in family meals more often, and employed the well-known Evelene Spencer, the "Fish Evangelist," to educate Canada's women on better fish cooking methods. In 1930, the department also provided modest financial support for advertising canned salmon in the Australian and Canadian markets. Furthermore, throughout the 1920s and '30s the department invested in National Fish Days and in regional and national exhibitions and demonstrations to showcase the science and economics of production as well as the economic and health value of increased fish consumption by Canadians.[1] In 1936, however, the department embarked on a much more systemic, organized, and better-funded marketing campaign lasting until 1940. The department continued its theme that only through increased domestic consumption could the fishery industry recover from its decades-long downturn.

This history of government-funded marketing to buttress a private-sector industry can be seen as a case study of the ways in which governments try to stimulate economic recovery. During the nation's most pronounced period of economic recession, 1920 to 1939, the federal government funded advertising and marketing campaigns to promote increased domestic consumption of fish and shellfish products specifically as a means to reverse a period of economic decline. The Department of Fisheries believed that increased domestic consumption could aid all levels of the fisheries industry, including fishermen, shore labourers, processors, wholesalers, distributors, and retailers. This focus on increasing domestic consumption, however, overshadowed other potentially more effective policies. The United Maritime Fishermen (UMF), a union of Atlantic Canadian fishermen, for example, protested the funding of advertising, arguing that it only helped trawler operators find a market for their sub-par catches that glutted the market and drove down the reputation and value of even superior-quality fish. Instead, the UMF called upon the Department of Fisheries to ban trawlers and *decrease* the supply of fish on the wholesale market so as to increase prices paid to fishermen for their catches. This labour-driven argument, however, lost out to the consumer-side argument as expressed by the leaders of the industry.

The department and industry focus on increased consumption ultimately proved ineffective. Domestic consumption did not increase substantially, and, following the wartime growth of the 1940s, the industry again fell into depression during the 1950s. Moreover, by focusing so much on trying to increase consumption, the Canadian government ignored production-side problems, including the poor quality of the product the industry put into the marketplace. During the first half of the twentieth century, government and industry leaders had both called for increased production by limiting restrictions, including the ban on trawlers called for by the leadership of the UMF. The long-term environmental and economic ramification of a consumer-oriented economic recovery plan are now apparent in the current state of fisheries collapse.

There was little debate about using government resources to promote increased seafood consumption in Canada during the world wars. These efforts fit neatly within larger policies regarding food rationing and nationalism. Once the First World War ended, however, debate emerged in both government and industry as to the rationale of such government involvement to promote the well-being of private-sector businesses. Advocates argued for the continuation of government financial support largely because they persisted in the belief that the fisheries remained tragically depressed. For them, the fisheries were not just another industry but were nationally important in terms of economic stability, international politics and trade, and cultural relevance.

In the 1920s, as the rest of the national economy was supposedly booming, natural-resource sectors suffered sharp decline compared to their wartime growth. Agriculture, forestry, mining, and fishing all experienced declining markets and reduction in both profits and wages. Fisheries decline became a primary focus for a 1928 Royal Commission and also received considerable attention from Henry Herbert Stevens's 1934 Price Spread Commission. Both of these high-profile investigations questioned the legacy of Canada's National Policy. Put briefly, the National Policy of the late nineteenth century had rested on government subsidization of extraction, processing, and exporting of natural resources, the connecting of regions through a series of transportation networks, and the encouragement of large-scale national corporations that policy-makers believed would bind the Canadian nation together more effectively than local or regional economic units. In sum, the National Policy prioritized the success of national industrial units based on resource extraction, processing, and exporting.

Yet despite this attention to national economic growth, severe regional underdevelopment continued. The poverty of the Maritime provinces became an important topic during the 1920s. The 1928 Royal Commission provides a rare look into the diverse levels of business operations within the Maritime fisheries economy. The justifications for calling the commission in the first place, the methods it used in its investigations, and the specific data it collected from witnesses are not germane to this study; however, what needs to be apparent in our understanding of the federal government's efforts to spur increased consumerism as a mechanism for industry recovery is that the commission conveyed a widespread belief that the Atlantic fisheries were tragically depressed, that the depression appeared to be universal across the industry, and that no one sector seemed to be gouging any other. Moreover, the industry must recover for national economic stability, and one of the best tools for this recovery effort would be government involvement. The commission's findings, as well as those of a Minister's Tour in 1929, largely focused on the need to create fishermen's cooperatives, invest more in science, secure better and more stable foreign markets, create a separate department of fisheries, and, most importantly for this study, encourage increased domestic consumption of seafood.[2]

The *Report of the Royal Commission Investigating the Fisheries of the Maritime Provinces and the Magdalen Islands* recorded the conclusions of the 1928 commission chaired by Justice Alexander Kenneth MacLean. The commission held 49 hearings, heard from 823 people, and accumulated over 5,700 pages of typed notes. It set out to find answers to nine specific questions, the first of which was "What should be done to increase the demand for fish both in the home and foreign markets?"[3] Although the report covered a variety of issues including the geographic distribution of fish and fishing activities, types of species caught, and methods of production and distribution of various fish products, the consumption of fish and limited demand for fish in the Canadian market remained dominant themes throughout the investigation and the report as it sought methods to improve the economic conditions of the fisheries industry and fishermen.

The commission's report began by noting market conditions. Because the weather for 1926 and the winter months of 1927 was very good, "the quantity of fish landed during that time was greater than the demands of the market and greater than in any corresponding period for many years previously." This was true for inshore fishermen and had also "enabled steam trawlers to land larger catches than usual." Yet the report further observed that "by reason of the low level of prices paid to fishermen in the spring and summer of 1927, particularly in eastern Nova Scotia, the fisherman was not receiving a reasonable return for his labour." Low wages were the result not of poor production but, on the contrary, good production: "It is quite clear, we think, that in the period mentioned there was an unusual quantity of fresh fish on the market, with the result that there was an excess of supply, and keen competitive selling." A glutted market had led to decreasing market value and collapsed fishermen's wages. It further concluded, "The market condition was not wholly attributed to overproduction by those normally engaged in the production of fresh fish." The salt-fish trade in 1926 was also depressed, leading many shore fishermen to abandon that trade and enter the fresh fish market. Cheap cod from Norway and Iceland having pushed Canadian salted hake out of the South American market, these shore fishermen pushed hake into the fresh-fish market in Canada. Traditional markets in the western United States had also been closed due to increased production in the American fisheries and so "turned into the Canadian market considerable quantities of fish that had been intended for the United States."[4]

This section of the report concluded: "Whatever may have been the causes contributing to this situation, and they were many, they resulted in price cutting, unprofitable alike to shippers and fishermen."[5] By stating that overproduction was not the sole contributor to the glutted market, the report signalled that overproduction by steam trawlers was not the sole contributor to the glutted market because shore fishermen also sold their catches in the fresh fish market. Nonetheless, from a consumer-economic perspective rather than a producer-economic perspective, the result was still overproduction beyond market demand. Thus, around the mid- to late 1920s, shore fishermen of Atlantic Canada increasingly came into competition with trawlers not just in the extraction and production of fish, which happened out at sea, but also in the selling of their products in the competitive consumer market.

During the commission's investigation, the Canadian Fisheries Association (CFA) called for the direct government aid to improve those market conditions. The association played a leading role in reinvigorating the idea that the government needed to facilitate increased domestic consumption of seafood. It justified its position by arguing not just that it would be good for the fisheries industry but also that it would positively affect individual health and national economic stability. The Atlantic fisheries, the commission observed, had a long history of direct government subsidies. In discussing the growth of the fresh and frozen fish industry, it noted, "Provisions having been made for air freezing fish, cold storage plants at the coast, refrigerated railway cars, and cold storage plants at points of consumption, it became possible to ship frozen fresh fish further inland and thus develop new and wider markets"[6] – and government subsidies had promoted all these developments. The commission's conclusion expressed the idealism, and even faith, of the industry that expanding the marketability of fish would solve the depression within the fisheries, but also reflected the belief that that marketing required government support. The commission specifically noted the success of government funding covering up to one-third of railroad express charges on less-than-full carload lots of fresh and frozen fish from Atlantic Canada to points west. The commission was unable to acquire data to prove that such support resulted in increased domestic consumption of fresh fish in central and western Canada; however, it assumed that since government payment for the program had increased by 300 per cent between 1909 and 1918, it must have succeeded in replacing American-caught fresh fish with a Canadian supply in important markets like Montreal and Toronto. It is a strange argument to make that increased government aid was a sign of improvement within an industry, but few people seem to have questioned it.

The CFA tried to sway the 1928 commission to endorse government funding for advertising to further supplement existing government aid in developing market access. The response from the commission was mixed. It was unanimous in its belief that "central Canadian markets are capable of absorbing a much larger quantity" of fish, particularly if rapid-frozen ground fish was better marketed. The commission regretted that the industry did not do more to facilitate increased consumption in the domestic market and found that "an unfortunate lack of co-operation" plagued many of the industry sectors, especially that of canned fish products. Despite persistent lobbying from the CFA, the commission concluded that "those who profit by such increased demand should therefore reasonably be expected to pay for the advertising. In the circumstances we are unable to recommend that the Department should at present grant a sum of money for this purpose."[7] Although this main governmental investigation into the problems of the fisheries endorsed the industry's assumption that increased consumption was the answer, it failed to conclude that such increased consumption was the government's responsibility. These conclusions fit squarely into the historical context of the 1920s, which would not change until the Great Depression forced the government into more direct involvement.

The assumption that increased domestic consumption was possible and would solve the perpetual economic woes of the nation's fisheries gained traction throughout the 1920s from various quarters in and out of government. For example, in 1923, MP Albert Frederick Healy of Windsor, Ontario – certainly not a dominant fishing centre – sent to Ernest Lapointe, then minister of fisheries (1921–24), a report from R. Sykes Muller, an advertising executive from Montreal, encouraging Lapointe to hire Muller for the new government-funded fish advertising campaign, which Healy mistakenly thought existed. One is left to wonder whether such action was due to Healy's honest concerns for the nation's fisheries or was done simply because Muller was a "personal friend."[8] Nevertheless, Muller's report provides an early view of the situation from the new occupational class of marketers – "ad-men" of the 1920s. Muller articulated the idea that advertising was advertising regardless of the product, and that "there is no reason why promotion to increase consumption of Canadian fish should not follow demonstrated merchandising methods." In fact, Muller completely dismissed the idea that the fisheries were unique; he was not in the least concerned with production rates or methods but only with "suggestions to create greater sales volume." For Muller, and presumably other advertisers, the solution to the fisheries problem was a matter entirely of consumption. Muller was confident that, with his help, industry could increase sales by 25 to 50 per cent, and placed considerable emphasis on the need to focus on canned seafood to achieve this goal. By focusing on canned seafood, he maintained he could create a year-round market for Canada's fisheries products: "There is no reason why fish should not be eaten just as freely in Winter as it is in the warmer months."[9] His words show a misperception that canning seafood, and theoretically all foods, freed them from environmental limitations, demonstrating his ignorance of the fact that seafood was very much tied to environmental realities. In fact, winter was actually the peak season for seafood sales, whereas summer sales lagged due to lack of artificial refrigeration among many smaller retailers.

Although Muller suggested targeting both consumers and dealers, he placed most emphasis on direct-consumer campaigns, using daily and weekly newspapers, free cookbooks, and school activities for children. Consumer education was key: "We do not feel that the average person has ever had sufficient knowledge regarding our fisheries and few persons know the different kinds of fish when they see them and fewer still know the different kinds of fish most suitable for a given purpose." Muller called for a national fish week, poster stamps and circulars, dealer cards, and newspaper advertisements, all marketing the slogan "Eat Fish Once a Day for Health and Economy."[10] Although his report exposed his ignorance of the fisheries economy, it did capture a key problem that it faced: few consumers knew much about fish and other seafoods. Marketing had the potential to increase product knowledge and sales.

By the time the Royal Commission met in 1928, marketing had gained traction as a possible solution to the poverty of Atlantic Canada's fishing communities, and the CFA had already embarked on what it hoped would be a three-year campaign to build upon the success of the Food Board's wartime efforts. This industry cooperative planned to raise $10,000 from its members each year, and called upon the government to provide an additional $30,000 annually. Their strongest ally within the Department of Fisheries was John Cowie, its acting assistant deputy minister. Originally from Scotland, Cowie had already played a major role in getting the department to consider consumerism in its policies. It was Cowie, for example, who was primarily responsible for adding a restaurant to the Toronto Fisheries Exhibition.[11] In September 1923, he met with industry leaders in Montreal and subsequently wrote an internal departmental memorandum: "I am heartily in sympathy with the efforts of the Association in expanding the sale of fish at home in this way, especially at this time when our fisheries are suffering considerably as a result of the Fordney-McComber Tariff in the United States."[12] Although he and the CFA did not get the $30,000 they wanted, the Privy Council did approve $10,000 to promote domestic consumption of the nation's seafood.[13] The Department of Fisheries appointed Cowie as its representative on the newly formed Canadian Co-operative Fish Publicity Fund (a marketing subsidiary of CFA), to facilitate industry-government cooperation in a six-month advertising campaign in the winter of 1923–24 with the hope of expanding sales during the industry's strongest retail season.

The 1923–24 campaign was decidedly a consumer-based program, making it distinct from how fish was traditionally marketed. The producers of the product, in this case as represented by the CFA and the Department of Fisheries, sidestepped wholesalers and retailers and went directly to consumers. This move created some animosity. B.T. Hutson, the editorial manager of Canadian Grocer, voiced concerns in a letter to Joseph Arthur Paulhus, president of the CFA, in November 1923. Hutson argued that Paulhus and the committee had marketed fish as an attractive product but had done nothing to ensure that grocers and other retailers could handle the new demand and provide high-quality product and service to the consumer. In fact, "the retail dealer has not even been apprised that there is a campaign on fish on. He has not been urged to make any preparation for it, and, of course, has not been shown how it can be handled and displayed so that it will reach the public in the proper manner."[14]

Hutson's letter had its effect. In January 1924 the Canadian Co-operative Fish Publicity Fund placed an advertisement in the *Canadian Grocer* specifically designed to encourage grocers and other retailers to push fish sales in their stores. It stated, "If you stand behind this campaign – give it your support and help to induce customers to 'Eat More Fish' – you help yourself, your customers, and your country." The ad highlighted the campaign's key tenets, which included economic, political, and social rewards for the nation. By selling more fish, grocers would help "one of Canada's greatest natural resources," provide "a decent livelihood for Canada's splendid fisher folks," encourage "greater prosperity for the Dominion and everyone in the Dominion," and, of course, generate "increasing profits."[15] The publicity fund also reached retailers and wholesalers via brochures, displays, signage, and other media, but the target was primarily consumers. The advertising content mainly focused on individual health. Questions of national pride and aiding the fishing communities were minor factors in this consumer-focused advertising, though they were key characteristics for the justification of the government funding.

Assessing the success or failure of this campaign is difficult. There was apparently no hard evidence that Canadians consumed more fish on either an individual or a national-average basis during or immediately following the six-month advertising blitz.[16] Paulhus claimed success largely based on the fact that over 75,000 people had requested the free recipe booklet offered in the ads. This might demonstrate a desire on the part of Canadians to better understand fish and learn more ways to prepare it, but it does not indicate that Canadians increased their consumption.[17] *Canadian Fisherman*, the trade journal of the CFA, not surprisingly also proclaimed the campaign a success: "On every side evidence has accumulated during the summer months which clearly demonstrates the effectiveness of this propaganda. No branch of the industry dependent upon domestic outlet has failed to feel the effects."[18] Yet it provided no evidence of increased sales and, although one might assume it was a knowledge source, it was hardly an unbiased one. The *Halifax Herald* also seemed to assume that increased advertising would naturally result in increased consumption and help in the recovery of the industry. A December 1925 story claimed that sales of Nova Scotia fish had tripled in the previous three years. It offered no statistics or evidence, however, or any basis for the idea that an increase in sales over a three-year period was related to a very recent six-month advertising campaign. The *Herald* noted that although much the advertising content focused on the nutritious value of fish, its real value was the "promotion of one of our principal industries."[19] Although evidence of success was limited and circumstantial if not completely unreliable, the CFA and the industry's media allies saw advertising as the solution to the depressed state of the industry, and in the winter of 1924–25, the association again asked for more federal support.

The original $10,000 that the Department of Fisheries had provided was for a six-month campaign in 1923–24. In February 1924, James H. Conlon, chairman of the Canadian Co-operative Fish Publicity Fund, wrote to William A. Found, director of the fisheries within the Department of Marine and Fisheries, assuming that if the department had provided $10,000 for a six-month campaign, it would surely provide $20,000 for a year-round one. As appreciative as Conlon would be for that amount, he actually wanted $50,000. To assure Found that the CFA did not seek the support merely for the benefit of the major producers, distributors, and larger retail outlets who made up its membership but had in mind the best interest of the entire industry from labourers on up, he included in his letter a resolve from Gloucester County, Nova Scotia, adopted the previous December, stating that the "livelihood [of] thousands of our population" rested on increasing domestic consumption of fish.[20] Conlon did not identify the authors of the resolve, who were likely the town's businessmen or board of trade rather than, or in addition to, the community's fishing population.

It was the responsibility of CFA president Paulhus to convince the Department of Marine and Fisheries to continue funding the industry's advertising. In a lengthy letter of 2 June 1925, to Alexander Johnston, a deputy minister of marine and fisheries, he outlined the importance of the industry to Canada, the value of advertising to the stability and growth of the industry, and the industry's inability to fund such advertising itself. He projected that if the department could contribute $50,000 each year for the next three years – a "paltry sum" – then the industry would grow four times over, eventually doing $200 million of business annually. Such growth, according to Paulhus, would stimulate economic development in all sectors of the economy.[21]

Paulhus did not get a confirmation of additional funding until November, when the Privy Council authorized another $10,000 "in light of the evidence of the beneficial results that have already accrued to the industry from the advertising of the last two years."[22] The amount hardly met Paulhus's request, and furthermore, the Privy Council insisted that the funds would only be provided as matching funds to money raised by the CFA, placing the burden on the industry to contribute, even though Paulhus had articulated it was not in an economic position to do so.

Support of this government intervention was not as universally accepted within the industry as Conlon and Paulhus suggested. In August 1924, the *Imperial Food Journal and Empire Produce News* had carried a review of the campaign, arguing that the Canadian fisheries were still "in a comparatively crude state of development," and that instead of trying to address the lack of technological progress in the catching and processing of the product, the leaders of the industry had turned to government handouts. The British observers argued that even if demand was stimulated, it was unlikely that Canadian fisheries could provide enough high-quality product to meet it. Forcing poor-quality fish on consumers could instead have the long-term negative result of forever turning them away from fish. In hindsight, this prediction proved accurate: by the 1950s, the poor quality of seafood products would become the dominant consumer issue for the Department of Fisheries. Although the CFA largely ignored the focus on improving quality, fishermen's unions, like the United Maritime Fishermen, occasionally voiced such concerns. For the time being, however, those most focused on consumerism seemed mainly interested in securing financial support for a domestic advertising campaign.

Fractures had also appeared within the business leadership of the Canadian fishing industry. In December 1924, G.H. Langtry, secretary of the Nova Scotia Fisheries Association of Yarmouth, sent a letter of protest to Pierre-Joseph-Arthur (Arthur) Cardin, the new minister of marine and fisheries,[23] stating that "the Association does not consider it good business for any firm or corporation to expect the Department of Marine and Fisheries to assist in its advertising propaganda."[24] That objection was nearly damning to the CFA's effort to secure government funding. Deputy Minister Alexander Johnston immediately wired the new CFA president, Arthur Boutilier: "In view of nature of this protest and interests represented [the] Department considers [it] inadvisable [to] proceed further until [the] matter [is] more fully considered."[25] However, something must have transpired behind the scenes, as the president of the same Nova Scotia Fisheries Association, Austin E. Nickerson, sent a telegram to Johnston insisting that Langtry did not represent the organization and that his opinions concerning the advertising were his alone. Langtry then wired Johnston informing him that he had been mistaken in his initial letter and his position should not be understood as representing that of the organization. Johnston was happy, therefore, to reassure the CFA that the money would in fact be provided. Nonetheless, this brief exchange exposes differences in opinion among industry leaders.[26]

Wholesalers also expressed concerns about the campaign's timing. F.T. James of Toronto, one of the largest metropolitan distributors of seafood in Canada, wrote to B.K. Snow, the new chair of the CFA's Publicity Committee, that he was withdrawing his contributions to the fund because he objected to advertising fish in the summer when there was generally little demand.[27] Snow replied that limited demand during the warmer months was precisely the reason advertising was needed to provide consumers with a "continuous reminder that fish – a good food, an ideal summer food, a health food is available – always, not only during the winter period."[28] Snow did not address the lack of proper refrigeration among retailers that made summer sales of seafood difficult. The most convincing evidence that the association did not have the full support of its members, however, was the limited contributions that industry leaders made to the campaign. The federal government had pledged $10,000 in matching funds, but by the fall of 1926 (the industry had missed out entirely on the 1925 season due to lack of organization), the CFA had only raised $4,942.[29]

Within the Department of Marine and Fisheries, John Cowie vocalized the opinion that the department should pick up the slack and contribute $15,000. William Found, who as the department's deputy minister was primarily responsible for coordinating the cooperation between the department and the CFA, disagreed. Found wrote to J.T. O'Connor, the new president of the CFA in 1926, that industry partners needed to recognize for themselves the importance of advertising, and that there should be no difficulty in raising contributions from individual businesses; furthermore, should the cooperative fail to do so, "it should not expect that the Department should undertake to make up the difference."[30] Support for a government-funded and consumer-focused revitalization of this private industry was failing.

Some of the most cutting and vocal opposition to this program, however, came not from industry leaders but from fishermen. Hilaire

Samson, president, and Andrew Sampson, secretary, of Local 2 (Petit Degrat/Richmond) of the Fishermen's Federation of Nova Scotia provided the most articulate of the fishermen's opposition to the program. Written in French, their January 1928 petition to Cardin succinctly stated "such a grant of public money would be useless to the fishermen."[31] Their objection was placed within the context of the larger debate on steam trawlers in Nova Scotia. Fishermen's unions had long voiced strong opposition to trawlers and would continue to do so throughout the 1930s and '40s. The union argued that the consumer market was depressed not because of limited demand but because of a glut that was the result of, and compounded by, low-grade fish produced by trawlers. Once consumers tried the poor-quality fish, they would be discouraged from ever eating seafood. In the view of Samson and Sampson, it was better to provide consumers with higher-quality fish caught by line fishermen even if this meant lower supply levels and more seasonality in fish consumption. Their argument is a good example of how workers who were directly engaged with the environment and observant of market trends had a better grasp of the economic and physical realities than bureaucrats and business elites.

The petition also articulated a class division between the capitalists within CFA and the workers within the union: "The trawlers operate only for the welfare of the holders of shares in the Companies, that is to say a few wealthy men compared with thousands of fishermen who with their families endure all kinds of sufferings and are compelled to leave the country on account of these trawlers."[32] The petition claimed the trawlers were "the mortal foes of the fishermen."[33] The 1928 Royal Commission gave these fishermen an opportunity to vocalize their opposition to trawlers, which incorporated a critique on the poor quality of the product yielded by trawlers and the impact it would have on the consumer market and the price or value of all fish.

The broader political issue of the "trawler question" – whether to ban steam trawlers in Canadian waters – set the context for the debate over the marketability of trawler-caught fish. "Trawler" more accurately defines the method of catching fish than the vessel itself. Put simply, a trawl is essentially a bag net, held open by a beam across its mouth or otter boards acting as wings, dragged along the bottom and scoping up nearly everything in its path. The method was first used centuries before by sailing smacks and was immediately controversial. Historian Robb Robinson noted in his 1996 book *Trawling: The Rise and Fall of the British Trawl Fishery* that the first known petition against trawling was sent to Edward III in 1376, and the first recorded fine for illegally fishing with a trawl dates to 1491. In England, the government began "strict regulations" in the seventeenth century, and a 1714 order remained in effect for 120 years. Yet regulations seldom came with actual enforcement, and trawling continued unabated. The latter half of the eighteenth century witnessed significant improvement in trawling vessels, and by the second half of the nineteenth century, the catch of sailing trawl smacks surpassed that of line fishermen in the North Sea.[34] By the 1880s and '90s, steam-powered trawlers replaced the sailing smacks and dominated North Sea and Northeast Atlantic fishing through the early 1900s.

Throughout this period of growth, at no point did trawling go uncriticised. Directed at either sailing or steam trawlers, this criticism was surprisingly consistent throughout the long history of trawling around the North Atlantic. Critics, mainly line fishermen, almost always focused on three clearly articulated concerns. First, trawlers destroyed the bottoms, effecting the base of not just the ground fish that trawlers sought but also a wide variety of other marine species important to fisheries economics; second, they produced more fish than the market could easily absorb, driving down market value and depressing line fishermen's wages and livelihood; and third, they produced an inferior-grade fish that either had to be thrown overboard or, if brought to market, ruined the consumer appetite for all seafood. The second and third critiques dealt with markets. Trawling only worked if it could market its lower-quality product. Those fishermen who opposed government-funded marketing argued that it only helped trawlers because they dragged up unmarketable fish and thus relied on government subsidies to shore up their losses and make trawling "profitable."

Three broader historical trends had set the context for the success of trawlers. The first was steam-powered industrialization, which effected both production and sales of trawler-caught fish. The first steam trawlers in England were just paddle-wheel steam tugs with improvised trawling gear. Because they were tugs, not built for offshore weather, they had to stay close to shore. Their limited available storage, which had to be split between coal for fuel and fish for market, also meant that they could not go far from home and were often limited to a twenty-four-hour fishing trip. The first purpose-built steam trawler was the *Pioneer*, which ventured out of Scarborough,

England, in 1881. At ninety-four feet long, it dwarfed the smacks. Purpose-built steam trawlers were two to three times as expensive as sailing smacks and required a consolidation of capital not easily acquired by moderately incomed fishermen. Traditional trawling ports like Hull, Grimsby, and Scarborough all transitioned to steam-powered trawlers within a decade; Hull and Scarborough shipyards built their last sailing smacks in 1886 and Grimsby in 1893. When the otter trawl appeared in 1895, it proved far more efficient than the beam trawler but was unworkable with sailing vessels. The final trip of a sailing smack was out of Grimsby in 1903.[35]

Steam-powered industrialization also shaped the marketability of trawler-caught fish. Robinson articulates that trawling in England emerged as a powerful force only within the context of industrialization and urbanization. The expansion of railroad lines allowed fishermen to increase the geography of their market for fresh fish well beyond the shoreline, just as industrial-urbanization, and an increase in immigration from Ireland, created concentrated urban populations of low-wage workers and their families. Whereas the volume of production was somewhat limited with line fishing, trawlers seemed to catch as much as any urban market could demand. Yet trawlers produced a volume of low-grade fish during a time of demand for higher-quality fish. No longer satisfied with pickled, smoked, or salted fish, the plebeian palate now wanted fresh white fish. To reconcile increased demand for quality with the output of poor-quality product, fishmongers turned to deep-frying, which could conceal poor-quality fish. By 1912, there were 25,000 fried-fish shops in Britain, consuming at least a quarter of the nation's 800,000 tons of fish. Nearly all the fried fish came from trawlers. Railroads, urban centres, and fried fish thus allowed trawlers to market product previously deemed "trash" and to dramatically increase production, placing more stress on the environment.[36] The steam revolution provided trawlers with their own mode of transportation and railroads to transport their catch inland to markets. Steam also powered the industrialization of England, setting the foundation for a rising urban working class to consume fish.

The second important historical theme explaining the trawlers' success is science. Between the 1860s and '80s, fisheries science concluded that humans could never significantly affect the population of fishes. The conclusion, most often associated with Thomas Huxley and his 1866 Royal Commission and oft-cited 1883 speech at the

International Fisheries Exhibition, was rooted in the observation of the millions of eggs that fish produced every year and remained the main line of argument for fisheries science for decades. This faith in the reproductive potential of fish coincided with the third crucial context for the success of trawlers: the rise of laissez-faire liberalism. Emboldened by the idea that conservation was unnecessary and with new faith in free-market capitalism, governments around the North Atlantic rejected most efforts to regulate high-seas fisheries (and even some of those inshore).[37]

The British story was similar to that in the United States. Matthew McKenzie shows how the rise of Boston's fresh fish industry was directly connected to trawling. The fish trust that McKenzie describes in *Breaking the Banks* knew better than most fisheries historians the centrality of market economics. They continually sought to control not so much the catching of fish as the transfer of the catch from boat to pier, and the transformation of the natural product fish to the consumer product seafood. As in Great Britain, the introduction of trawling immediately generated opposition from line fishermen along the same three lines of critique (destroyed the bottoms, drove down the value of all fish and thus the wages of line fishermen, and produced a poor-quality product without a market). Beam trawling in the United States grew by fits and starts. The early efforts of the 1860s failed to generate a profit because of the catch's poor quality. McKenzie's study, like Robinson's on the English trawler fleet, shows how the integration of the waterfront into urban marketplaces provided an opportunity to dispose of a low-grade fish as food for the urban working class. Increased urbanization throughout the late nineteenth century and improved transportation lines between New England's fishing ports and northeastern cities provided the essential context for success. By the early 1900s, trawling dominated the fresh fish trade in Massachusetts.[38]

In Canada, fisheries leaders in both industry and government knew this history and by the 1920s sought to aid trawlers by improving marketability. The trawling industry had insisted that it was necessary if the industry hoped to improve market conditions, because only trawlers could provide the regular supply of fresh fish needed to ensure consumer confidence. Opponents argued that the low-quality fish undermined consumer confidence in all fish products and oversupplied an already glutted market. The 1928 Royal Commission gave voice to both sides of the debate.

The commission was unequally split over the "trawler question" and thereby submitted majority and minority reports related to trawlers in its appendix. The majority report seemed supportive of the shore fishermen's critique of the market impact of trawlers but largely dismissed concerns around the impact of trawlers on the environment. It focused mainly on the arguments that, first, trawlers marketed an "inferior product, which in the end injured the industry by discouraging the consumption of fish," and, second, the over-production caused by trawlers glutted the markets, preventing "the shore fisherman from disposing of his catch, of superior quality, at a reasonable price."[39] This report was particularly concerned with the second objection and concluded that the real issue was one of control. As the productivity of trawlers increased, those companies that purchased fish from fishermen – the biggest of whom also happened to own trawlers – would be less likely to purchase it from shore fishermen. Even if they did, they would offer low prices. In other words, despite recognizing the problem of overproduction, the commission did not seek to decrease supply but instead sought to redirect the control of that supply to give more selling power to shore fishermen.[40] "If steam-trawlers are used to full capacity," the majority report concluded, "there is, therefore, little promise of markets for the shore fishermen, under present conditions, for as consumption increases steam-trawlers will doubtless increase."[41] It is doubtful whether the majority report would have objected to increased production by shore fishermen if they had come to dominate the supply of fish.[42]

The minority report, written solely by Alexander Kenneth MacLean, exposed this contradiction. MacLean accused the majority report of being anti-modern and instead argued that the whole of the community would benefit if Atlantic Canada increased its productivity, something the majority report did not dispute. Steam trawlers, according to MacLean, provided increased production and increased opportunity for labourers both at sea and in shore-side production facilities, all while producing a superior commodity at a reliable rate.[43] In the end, neither the majority nor minority report suggested decreasing production or adhering to what appeared to be limited demand in Canada for fish. Their debate was really over who controlled the increased productive capacity of Canada's Atlantic fresh fisheries – trawlers or shore fishermen.

Fishermen unions continued to argue that trawler-caught fish was inferior and spoiled the market by driving down consumer interest.

At a 1927 meeting of fishermen in Canso, Nova Scotia, for example, William Oliver called trawler-caught fish "stuff not fit for food," while David Walsh called it "diseased" and "not fit for human consumption."[44] This opposition continued into the Depression years when, in 1931, the United Maritime Fishermen submitted a resolution to the Department of Fisheries that no government funds should be used to support the advertisement of fish without the cooperative's approval. The United Maritime Fishermen wanted the government to provide marketing assistance only for high-quality fish, which for them meant fish caught by line fishermen. Such fishermen opposition, perhaps not surprisingly, largely went unanswered by department bureaucrats.[45]

A 1938 meeting in Halifax to discuss the broader "trawler question" did address the question of the market impact of trawler fishing. The Maritime-National Limited Company of Halifax, one of only two trawler operators in Canada, insisted that trawlers were central to the development and maintenance of a fresh fish market in Canada. Market demands had clearly shown that the Canadian consumer was not interested in salted or processed fish but increasingly wanted fresh fish.[46] Only trawlers provided a "regular and steadiness" of landing necessary for reliability of supply. Trawlers were producing 60 per cent of the haddock for the fresh fish trade, while vessel fishermen produced 35 per cent; independent shore fishermen landed the remainder. "Neither inshore fishermen nor vessel fishermen singly or together," the company continued, "can maintain the steady and regular supply of fish necessary for the fresh fish trade." Instead of seeking increased regulation of fishing methods, the company insisted "the need of the industry, salt and fresh, is markets, and markets only."[47]

Meanwhile fishermen's unions continued to press the argument of trawler-caught fish's negative impact on the market. In their brief to the conference, Angus J. Walters and W. Lawrence Allen of the Fishermen's Federation of Nova Scotia, made up primarily of shore fishermen, disputed that "steam trawlers are necessary to preserve continuity of supply of fresh fish." The federation argued that, in March 1938 alone, trawlers had produced four to five thousand quintals of trash fish that had to be redirected away from the fresh fish market and were instead salted, which only drove down prices in the salt fish trade.[48]

Back in 1928, the CFA continued to ask for government funds for advertising, calling for the Department of Marine and Fisheries to contribute two-thirds of a $100,000 campaign in coordination with the industry.[49] In its 1928 annual meeting, the CFA passed a resolution stating that "an advertising campaign of this kind is not only an effort to build up a natural industry, but it is a propaganda of patriotism," and that "the building up and developing of the fishing industry is not an appeal to sectionalism in any sense of the word, but a development of the muscular sinew of our commercial life for the good of the whole body politic."[50] With growing opposition, however, and with the 1928 Royal Commission's investigation on the totality of the fisheries industry of the Maritime provinces looming, the department never did provide the funds pledged by the Privy Council in November 1925.[51] When the Depression hit in 1929, government funding was largely cut off. Government encouragement of fish consumption would have to come in piecemeal fashion. A primary focus of the department in the early 1930s was the Pacific salmon industry, particularly canned salmon. Because such a sizable share of that production was exported, and because of the collapse of those export markets due protective tariffs, the federal government felt it had some constitutional power to intervene on behalf of a private-sector industry.

An opportunity emerged in 1930 when the Department of Fisheries and the Department of Trade and Commerce both partnered with the British Columbia Packers Association to begin an advertising campaign to increase both domestic and foreign consumption of Canada's canned salmon. The Department of Fisheries contributed $25,000 and the Department of Trade and Commerce contributed $10,000, while the industry cooperative itself raised an additional $56,000.[52] The domestic campaign included both radio broadcasts and newspaper advertisements managed by the McConnell & Fergusson advertising company in Vancouver. The content hit the usual points of the health value of fish and the ease of use of canned seafood, but within the context of the increasing anxiety over the Depression, the advertising firm felt it essential to include "the angle of economy and food value." By "food value" it meant economic not nutritional value (that is, the cost of nutrition, fat, and calorie per pound of food).[53] The campaign also touched on nationalistic and romantic rhetoric, often celebrating the heroic Canadian fishermen and asking the consumer, "When the housewife exchanges her money for canned salmon at her grocer, purchasing that commodity as a delicious food, rich in nourishing properties, does she ever stop to think of the background of hardship, heroism and toil that lies back of that prosaic little can?"[54] This program to spur consumption

was widely celebrated in industry-supportive and regional newspapers as being "real action" during a time of economic crisis.[55] The
government followed up the campaign by providing an additional
$50,000 of matching funds in 1931. Throughout 1930 and 1931, the
nation's newspapers seemed happy to cooperate with the government-
industry advertising partnership by pairing paid advertising with editorial content on the importance of the industry, the steps taken by
the government to promote its economic well-being, the need for
Canadian consumers to use their purchasing power to support a
nationally important industry, and the health value of Canada's
canned salmon.[56]

Salmon was Canada's largest single fisheries commodity. When the
Department of Fisheries embarked on an effort to provide direct subsidies for the salmon industry not just in foreign markets but also
domestic ones, it sparked controversies across the nation's fisheries
industry and accusations of favouritism. Minister Cardin noted this in
an April 1930 letter to James Malcolm, minister of trade and commerce. The British Columbia Packers Association had been to Malcolm's office, encouraging him to use funds to advertise salmon in
foreign markets. Cardin, however, initially warned against this, noting
that "different sections of the fishing industry are not alike in their
interest"; although a campaign might help the salmon industry, the
dried fish industry, which also depended on its exports, could be
pushed out of a foreign market if consumers there chose salmon over
dried codfish.[57] Allan McLean, president of Connors Brothers, the
nation's only sardine producer and one of the largest non-salmon
seafood canneries in the country, complained directly to William
Found about the department's aid to British Columbia packers: "We
think we are entitled to the same consideration as the Canners on the
Pacific Coast." Found's reply emphasized that the funding was made
available to the BC packers because they had successfully formed an
industry cooperative representing more than just one or two major
businesses. Yet requests for advertising support from the CFA, a much
larger industry cooperative, had gone unheeded, bringing Found's
explanation into question.[58] It may have been more significant that
the British Columbia packers had raised $56,000 within the industry
to commit to this advertising venture and that the government was
providing only $35,000 in additional funding – an arrangement that
would have answered the government's long insistence that it provide
only matching funds of a limited volume.

By the late 1920s and early '30s, therefore, the department was getting increased pressure from across the fisheries economy to do something about low domestic consumption and collapsing foreign markets. The CFA continued its persistent calls for federal aid for advertising.[59] Many nations around the world were now erecting tough nationalistic economic policies that set high tariffs on imported goods. US tariffs were particularly harmful to the Canadian industry. For exporting countries like Canada that relied heavily on the ability to export a large volume of its natural resources, the new economic reality was catastrophic. Canada exported over 80 per cent of its fisheries production. Within the new global context, the CFA's calls for increased domestic consumption gained new traction. In a public address in Prince Rupert, Alfred H. Brittain, chairman of CFA's Publicity Committee, again voiced the organization's long-standing request that the government provide $100,000 in advertising funding. Brittain tied this desire to recent calls for domestic consumption of domestically produced goods that had gained some public attention within the context of the Depression and the rising sense of economic nationalism in Canada: "It may also be remembered that throughout the length and breadth of Canada the question of 'Produced in Canada' is one which is being brought very potently to the attention of the rank and file of the public, and the slogan of buy goods made in Canada, very easily and very effectively dovetails with a slogan of Buy Fish Produced in Canada, and the health qualities of fish as a diet."[60] CFA secretary R.W. Gould forwarded Brittain's address to Minister Cardin, with the hope that the department could pair publicity regarding the new minister's appointment with calls to increase fish consumption. In a 1930 lobbying message to the ministry, the CFA declared that increased domestic consumption spurred by government-funded advertising would "almost completely eradicate the difficulties now presenting themselves to the industry."[61] That unquestioned faith in advertising as the single solution to the economic woes of the fisheries never seems to have subsided.

The CFA had some allies in government and the press. Ontario's maritime trade commissioner R.W.E. Burnaby[62] gave a speech in December 1930 calling upon the federal government to invest in advertising fish, stressing the social value of consumer purchases and industry well-being, insisting "an extensive campaign throughout the press at this time would stimulate the market and materially increase the consumption. Public interest in the welfare of fishermen and their families, the hazards of their calling, the hardships privations and handicaps under which they labor might well be aroused."[63] Despite the CFA's persistent pressure on the minister's office, however, the department took little action on direct consumer advertising via media outlets throughout 1929 and 1930.[64]

One of the easiest ways for the fisheries department to encourage increased consumption of fish domestically was to expand its use of radio. In his initial inquiry regarding what would become Mildred Campbell's radio talks of 1930, Found wrote to the Canadian National Railway Broadcasting Station, "Our objective of course is to aid in the further development of an important Canadian natural resource."[65] The department had already been using the radio to communicate pertinent and educational information to fishermen, but, in 1930, it began to target housewives for the purpose of "educating" them on the value of consuming more seafood. Although occasionally these radio talks included content on the national importance of the fisheries industry, its history and technological progress, and the need to improve the economic conditions of the industry, their content overwhelmingly focused on the Canadian housewife's need to provided her family with well-balanced, varied, and nutritious meals. In this regard, the Department of Fisheries seemed convinced that the problem of low consumption was rooted in the inability or unwillingness of Canadian women to cook and serve fish more often. Economic recovery via increased consumption therefore rested on a nationwide "educational" campaign targeting Canadian housewives, as discussed in the previous chapter. The strategy was appealing because it was simple, had a refined targeted audience, and would be relatively inexpensive in comparison to the mass advertising still being advocated by the Canadian Fisheries Association.

In 1932, Found himself went on the air as part of the Professional Institute of the Civil Service of Canada broadcast to provide insight into the Department of Fisheries' work. After outlining the usual talking points regarding the scope, size, diversity, and work involved in Canada's fisheries, as well as the health benefits of fish, Found expressed concern about the low domestic consumption rate – only twenty pounds per capita – and claimed that by adding "only a few pounds per capita per annum, many of the problems that are now facing the fishing industry would disappear." Consumers, particularly housewives, were increasingly being told that economic recovery for the fisheries rested in their ability to consume more.[66] Yet because the industry was so fractured by region, fish species, and product type, it was difficult to create a unified marketing voice. This is one reason why the leadership of the CFA concluded that only the national government could provide this universal messaging.

Early in the Depression, the Department of Fisheries did seek some outside professional consultation to determine how it could be more proactive in supporting the fisheries industry. In 1930, the Privy Council approved $12,000 in funding to hire the Montreal firm of Cockfield, Brown to do a comprehensive study of "the conditions under which fish are marketed and merchandized in Canada." These conditions would include consumer tastes, distribution networks, wholesale and retail operations, and current marketing activities.[67] The company conducted surveys in more than one hundred cities and towns in Canada; it sent out questionnaires to 3,000 customers, 700 restaurant and hotel managers, 700 retailers of canned foods, 500 retailers of fresh fish, 250 wholesalers of canned fish, 350 wholesalers of fresh fish, and 250 producers and shippers. The firm also sought the input of the production side of the industry by mailing questionnaires to 400 shore fishermen's association and union members, 400 vessel owners, 200 salmon canners, 500 lobster canners, 150 other fish canners, 400 fish-curing establishments, 200 fish reduction plants, 500 inspectors, and 125 transportation workers.[68]

It is not surprising that the investigators from Cockfield, Brown, a marketing firm, focused on the idea that the success or failure of the industry rested on the ability to sell the product in a food market. "The prosperity of Canadian fishermen," the company's report noted, "depends not only upon an adequate catch, but also upon the profitable disposal of the fish when caught." The core problem, however, was that while Canada was a very high producer of fish, at 120 pounds per capita, it consumed just over 20 pounds per capita. By comparison, the United Kingdom consumed over 40 pounds per capita, Norway 70 pounds, and Germany 21 pounds. Thus, the industry in Canada relied heavily on export markets, which accounted for the distribution of 70 per cent of its total value and 80 per cent of its total volume. Yet in the new era of protective economic nationalism around the world, such reliance on export markets was dangerous. The Canadian fisheries industry would be better served, the report concluded, if it secured a larger share of the domestic food market, in which "the producer and consumer are more closely in touch with each other" and the government could ensure "freedom from outside competition," meaning protective tariffs.[69]

The surveys sent out by the firm provide some important data on consumer preferences. They indicated that on average Canadians ate fish about twice a week. There were regional differences, with the Maritime provinces at the high end at 11.9 seafood meals per month and the Prairie provinces at the low end with 7.9 per month. Yet, surprising to many at the time, the surveys found few variations in the amounts consumed by different economic classes. The firm concluded that the key variable in increasing fish consumption in Canada was location. Those communities close to fishing centres consumed more fish.[70] They did so, the report further concluded, not because of cultural ties to the maritime environment and its resources but because of regular, convenient, and inexpensive access to fresh fish: "Per capita consumption is not so much effected by distance as by constant supply of suitable fish." This reading of the data, the report suggested, would have to conclude that the future of the nation's fishing industry rested on the fresh fish trade and getting more high-quality fresh fish to more Canadian consumers more quickly: "If a system can be established whereby truly fresh fish is constantly available at reasonable prices and consistently made known to the public as such, then the problem of consumption is a long way towards solution. It is such a system which must be found for the Canadian industry."[71] That conclusion seemed to support claims made by organizations like the United Maritime Fishermen and other fishermen labourers' organizations that the real issue was low-quality products from the trawler fleet. Yet steam trawler operators could counter, and often did, that only they could provide the regularity of landing demanded by the market.

Cockfield, Brown's conclusion, however, seems contrary to consumer responses to questions about why they did *not* eat more fish. Only 11 per cent stated that it was too difficult to get fresh fish. Another 11 per cent stated that it was too expensive. About 33 per cent of those surveyed replied that they felt that they ate enough fish to maintain a healthy and balanced diet; 20 per cent responded that they simply did not like fish.[72] Thus, more than half of the potential consumer base was not inclined to increase their fish consumption because they already believed they ate enough fish or simply did not like it. Factoring in the chance that even one family member's dislike of fish could limit its consumption by the entire family, the number of people unlikely to eat fish would surely be even higher. Looking at these numbers, any increase in domestic fish consumption rested not on improving access to fresh fish (only 11 per cent said that was a factor) but on convincing more people that they needed more fish for health than they previously thought or convincing more people that they actually did like fish, possibly by improving the quality of the product. Remember, nearly everyone involved in the fisheries – trawler operators, shore fishermen, unions, government bureaucrats, and trade organizations – argued that improvement of the industry rested on getting more and better fish to the consumer, and this report too maintained that this was the main issue. Yet the data simply did not support this conclusion. No one seemed to notice that only 11 percent of those surveyed expressed any desire for access to a better or more reliable fish product.

Yet this reading of the data was not the one adopted by Cockfield, Brown or by the Department of Fisheries. Instead, the remainder of the report, and the sections most often referred to by the department, focused on improving the distribution of fresh fish. Thus, the keys to industry expansion were seen as the more efficient and reliable high-volume catching of fish ideal for the fresh fish trade (trawler-caught haddock), integration of production and distribution capacity to provide improved efficiency, and more investment in modern refrigeration and sales technology at the retail level. These conclusions inevitably led the advertising firm into a discussion on the "trawler question." Although not part of its original investigation, it concluded that "the primary interest of increasing and improving the marketing of fish does, however, demand the greatest freedom in applying the most effective and, above all, regular means of securing the catch."[73] The report could not have more clearly given its support to trawler operators for their ability to increase production and regularity of availability of the product most in demand: fresh fish.

At the retail level, the firm's report emphasized improved refrigeration and adoption of brand-name marketing for fresh fish: "Modern merchandising practice endorses the use of brand products as they tend to standardize quality, and enable consumers to purchase on a sounder basis." When it came to canned salmon, where there were no less than 155 different brand names, the report suggested consolidation of the industry into fewer, more highly capitalized operations; the number of brands had become so confusing that consumers had a difficult time developing the brand loyalty that would facilitate increased purchases.[74] Furthermore, retailers had to be conscious of the quality and attractiveness of the product; quality-control inspection of fish needed to be expanded to include all levels of production, distribution, and sales.[75] The report suggested looking to Great Britain as a model. British fishing operations there had achieved successful horizontal and vertical integration, with efficient production, distribution, and sales of seafood that maintained high per-capita consumption. The Department of Fisheries, however, did not begin widespread quality-control inspection of the product until the 1950s.

The key theme throughout Cockfield, Brown's report was improving product availability (more so than quality) and more consistent marketing on a national scale. If a dependable product was provided on a regular basis, more people would consume fish. Yet to achieve this, the report mainly emphasized the advantages of consolidation at all levels of production, as in Great Britain. This, the report concluded, had to precede large-scale advertising.[76] The "disjointed conditions" of the Canadian fisheries was not a new problem but rested on the historical regionalism of production. In the new consumer age, such localism was a handicap in economic expansion; the fractured nature of the industry had to be brought into some "unanimity." Once integrated, it would have the capital necessary to invest in modernization that would drastically increase the availability of fish products and thus domestic sales. The federal government could aid in this process by providing some economic stimulus or reward for such integration. Given the constitutional limits imposed by the British North America Act, such direct government support of industry was potentially problematic, but the report suggested that "advertising on the part of the Government can be justified as providing the impetus to a well-defined policy of rehabilitation of that industry in the general public interest."[77] Getting consumers to eat more fish was in the national interest. However, government policy designed for economic stimulus was limited by the prevailing political faith in classical economic liberalism. The Bennett and Mackenzie King administrations both largely opposed government intervention for economic recovery and instead focused almost wholly on tariff and banking debates.

Although much of the social history – that is, the lived experience – of the Great Depression in Canada was similar to that of the United States, the political response at the federal level was vastly different. The British North America Act certainly created significant barriers to direct government aid in such a way that prevented a New-Deal-like response. Yet this factor should not be overemphasized. The political thinking of both Bennett and Mackenzie King limited their outlook to continual debates on classical political-economic issues like tariffs, balanced budgets, and currency regulation.[78]

Bennett had two main plans for addressing the Depression. The first was the Unemployment Relief Act, a $20 million aid package that eventually totalled $100 million. But the administration had no plan for effective distribution of the aid, and in the end the money served largely as basic subsistence relief that had little impact on economic recovery. Bennett's second strategy was the 1930 tariff, which effectively established economic nationalism, something Canada's fisheries were fighting overseas. Such a closed-door policy in Canada had the potential of sparking retaliatory legislation, or at least limiting diplomatic efforts to open foreign markets to Canadian fish. In 1932, Bennett tried to address this issue by hosting the Economic Conference in Ottawa, which sought to establish a British Commonwealth preferential trade network that could have gained Canadian fish entry into important foreign markets like Great Britain and Australia but did little to help efforts to reduce tariffs in their main marketplace, the United States. Thus, history most remembers Bennett for his infamous work camps and widespread protests against government suppression of oppositional voices. His 1935 "New Deal" speeches did little to save his political life, and the Conservatives retained only thirty-nine seats in the 1935 elections.

In 1935, Mackenzie King might have helped the Liberals win 171 of 245 seats in Parliament, but he was no Roosevelt. In fact, he strongly opposed the activist government in the United States, the New Deal, and, eventually, Keynesian economics. His interpretation of the Depression was most influenced by his comparison of it to the 1920 recession, which he believed he had solved the last time he was prime minster via tariff reform and a balanced budget. Within three weeks of taking office, he signed a freer trade agreement with the United States and in 1938 expanded it to include Great Britain. King expanded corporate and sales taxes and amended the constitution of the Bank of Canada, which Bennett had set up, to give the federal government more control over monetary policy. King also established an independent National Employment Commission to investigate relief payment, but in 1937 he objected to its proposal to increase direct government aid and stimulus for economic recovery. The budget debate of 1938 pitted him against his own minister of labour, Norman Rogers, over public works funding. In the end, the government provided a modest stimulus package of $25 million but beyond that did little to influence economic recovery directly. As can be seen, both Bennett and King were largely hostile to using the federal government or its ministries to provide direct aid or economic stimulus to private industry or to extend unemployment or poverty relief beyond its existing minimal levels.

In February 1934, the Bennett government established the Stevens Royal Price Spread Commission, chaired by Henry Herbert Stevens, Bennett's minister of trade and commerce. Stevens broke with the pattern of economic conservativism and used the commission to probe deeply into inequalities of Canadian society rooted in industrialization. Although historians have mostly used the Stevens commission to expose wage inequality within the nation's manufacturing and retailing industry, Stevens's investigation also explored the resource-based sectors of Canada's national economy. The commission came to some interesting conclusions regarding the balance of economic power within the nation's fisheries. Many of these conclusions countered the persistent calls of the Canadian Fisheries Association for a consumer-oriented solution to the industry's economic doldrums. In the end, even as the commission called for the government to secure better wages for the nation's fishermen via minimum price control, marketing boards, and labour cooperatives, the Department of Fisheries throughout the remainder of the 1930s and into the '40s continued its focus on increasing consumer demand.

Although Bennett himself had little interest in Stevens's Price Spread Commission and had not expected much from it, and furthermore was upset with the findings when they were released, the commission itself used the opportunity to lay down sharp, direct, and specific critiques of the inequalities of industrial capitalism in Canada. Its expressed intent was to "investigate the causes of the large spread between the prices received for commodities by the producer thereof, and the price paid by the consumer thereof," with the objective of outlining the steps the government might take to provide an "opportunity for fair returns to producers."[79] Although the fisheries were not specifically named as a subject, food was, and through the course of parliamentary debate it was concluded that fish was food for the purpose of the investigation; and so the commission probed into the inequities of wages and prices but limited its focus to the Atlantic fisheries.

Leonard William Fraser wrote the final report on Atlantic Canadian fisheries for the commission. It did not shy away from casting a large, dark shadow over the nation's fisheries, stating in the opening paragraph, "The information secured will place before the committee the serious conditions that exist in the fishing industry – conditions that are adverse to the welfare of the fishermen and to the advancement and development of the industry as a whole."[80] Fraser began with the 1927 Royal Commission report and the report on marketing Canadian fish produced by Cockfield, Brown in 1932. He concluded that, based on the latest industry statistics, "the industry has not improved since 1927, with the result that the present situation and certainly the position of the individual fishermen, is less favourable than was the case at the time of the investigation by the Royal Commission." The reason for this continued decline, he argued, was that although the fisheries had suffered from the most recent Depression, in reality, the industry had long-term, historic structural flaws that limited economic development even when the rest of the nation was "enjoying periods of rapid expansion."[81] Although focused on the fisheries economy of Nova Scotia, New Brunswick, Prince Edward Island, and the Magdalen Islands of Quebec, Fraser's report noted persistent declines in all economic variable between 1926 and 1933. He concluded that the main problem was "lack of coordination" within the industry.

Fraser's report set the fishermen, who, he argued "must always be recognized as the primary producer in this industry," within a larger economic structure that traced the commodity's value backwards from what consumers paid for fish to what fishermen received for their catches. Production inconsistency and poor product quality, as well as cultural barriers that limited regular consumption of fish throughout the week, months, and seasons, all led to wide and speculative price fluctuations at the retail level. Fraser concluded, "Price cutting and unfair competition is prevalent in all branches of the fishing industry." Fluctuations created too many economic variables and resulted in aggressive bidding wars that purposefully or indirectly drove down prices paid to fishermen. The fishermen's persistent economic depression throughout the 1920s and '30s resulted from their near total lack of bargaining power within an extremely volatile market, compounded by the intense instability of the national economy during the 1930s.[82]

Fraser essentially presented two conclusions, one a consumer-focused critique of the market and the other a producer-focused proposal to aid fishermen.[83] The first argued that although increased domestic consumption of fish, something long championed by the CFA, might seem an easy fix to the problem, any efforts to convince Canadians to eat more fish must be preceded by a dramatic improvement in product quality and dependability: "High quality of the fish available is a most essential premise of any program to extend the Canadian per capita consumption of fish." The quality of fish on the retail market in Ontario and Quebec was "not sufficiently high or uniform to lead to any considerable extension in the volume of fish sold." Fishermen's unions had already voiced concerns about poor-quality fish and specifically argued that trawlers were mainly responsible for trash fish flooding the market. Fraser did not mention trawlers but instead blamed the poor quality on retail merchants for not properly handling, displaying, or storing fish. Increased fish consumption as a means to improve the fisheries economy was thus "largely dependent upon satisfactory handling conditions"; there could be no positive impact on the economic well-being of the nation's fisheries before the industry or the government addressed poor retail and wholesale distribution and handling methods.[84]

Fraser's second conclusion dealt with the production side of the industry. Here he focused almost exclusively on the fishermen. The root cause of their depressed state was their lack of bargaining power and near total lack of any kind of control over the market. Like many before him, he recommended organizing the fishermen into cooperatives and establishing a marketing board to regulate the industry and set minimum prices to be paid to fishermen for their catch.[85] Yet cooperatives were already operating with limited success under the St Francis Xavier University extension program, and the Privy Council would later declare a fisheries marketing board unconstitutional. Thus, despite the fact that the initiative behind the Stevens Price Spread Commission was to secure producers better wages for their labour, nothing really emerged from it that directly benefited Canada's fishing labourers. The report produced from the investigation, however, provided an excellent in-depth assessment of the links between production of quality products and consumption levels.

Although Fraser gave approximately even weight to consumption- and production-based problems, and perhaps even leaned a bit towards the former, reviews of the report overwhelmingly focused on his analysis of the depressed state of fishing labourers rather than his critique of poor quality of the product on the retail market. Perhaps

it was the context of the Depression, perhaps it was that poor fishermen made better newspaper copy than market trends – whatever the
cause, the result was an immediate focus on the depressed state of
Canada's fishermen. Yet this focus did not lead to the adoption
of the suggestions made by those fishermen, namely, the banning of
trawlers in Atlantic Canada. Instead, an economic trickle-down proposal emerged from the reviews of Fraser's report, even if such an idea
was not in his report, that shaped fisheries policies for the next two
decades. Government and industry leaders both concluded that the
way to address fishermen's low income was to increase consumer purchases of their product, which would increase the volume of sales and
thus overall incomes. Neither government nor industry leaders
seemed to take any notice of Fraser's insistence that improved product quality had to precede campaigns to increase consumer demand.

John Cowie of the Department of Fisheries was one of the first to
review Fraser's report when he was asked by William Found, who
himself had been asked by Lester B. Pearson, secretary of the Price
Spread Commission, to provide an official reaction from the Department of Fisheries. Cowie concluded that Fraser's report showed
"clearly that the fisherman is not getting his due share" because when
markets collapsed at retail level, the burden of lost profits was always
shifted back to the fisherman, "the first and chief sufferer." Fraser
never quite said as much in his report, and never specifically said that
there was an unjust spread between the prices paid by consumers and
the wages paid to the workers. That did not stop Cowie, however,
from focusing on market demand as the source of the problem, a view
he had long expressed. Cowie proposed creating policies or systems
that would reduce markups at the retail level, which would stimulate
consumer demand by lowering prices and thus enable wholesalers
and distributors to pay higher prices to fishermen. Fraser had never
mentioned high markups as a cause of low demand; instead, this was
an idea long floated by the CFA and was the likely source of Cowie's
conclusions.[86] Despite Cowie's long history in fisheries and his deep
knowledge of all levels of production, distribution, and sales, one has
to wonder how much of Fraser's report he took seriously, as most of
what he wrote in his review seemed largely grounded in his own past
experience and his work with the Canadian Fisheries Association. In
other words, Cowie completely misrepresented Fraser's report. Yet
Cowie's was the official review presented by the Department of Fisheries, and it was Cowie's conclusions, not those of Fraser, that would

most shape the department's commitment to a consumer-focused recovery plan resting largely on increasing retail volume rather than retail quality.

The CFA published its own review of Fraser's report in the *Canadian Fisherman*, which, contrary to its name, was the CFA's own trade journal and did not reflect the views of the fishermen themselves. The review criticized Fraser's report for providing nothing new and its ignorance of "the economic axiom that when an over-supply of any commodity exists, there is little or no bargaining power" among the producers.[87] This general conclusion of the CFA is surprising for two reasons. First, Fraser did address market trends, although he focused more on the need to improve the quality of the product appearing on the retail market than on quantity of supply or demand. It is possible that Fraser's critique of the poor quality of retail fish product upset the retailers and wholesaler wing of the CFA, and so the organization felt it necessary to address Fraser's accusations. Second, although the CFA criticized Fraser for failing to recognize the "economic axiom" of supply and demand, its own solution to the over-supply problem was simply to increase the demand side of the equation.

Fishermen like those of the United Maritime Fishermen had long voiced concern over steam trawlers dumping trash fish on a glutted market and sought the government's assistance to ban such practices and reduce the supply side of the economic equation. While that one line about the "economic axiom" from the CFA's trade journal seemed to hint at some shared understanding of the problem, this was misleading. Producers within the CFA did not support the call to ban trawlers; their solution to the imbalance of supply and demand was to increase demand. The article continued, "The Commission has seen fit to ignore the recommendations of the Canadian Fisheries Association in the matter of a campaign to increase the domestic market for fish." The article concluded optimistically that "a campaign of educating the consumer in Canada" would result in increased sales of more than 100 million pounds of fish.[88]

The CFA wasted no time in shaping the rhetoric emerging after the Price Spread Commission. Alfred H. Brittain, chairman of the CFA's Publicity Committee, went on a lecture tour that received some national media attention. He continued to voice the CFA's call for $100,000 of government funding for a marketing campaign to increase domestic fish consumption but did so with increased emphasis on the argument that the strategy would provide aid to impoverished fishermen.[89] In essence, Brittain used the image of poor fishermen to call for a program that most poor fishermen opposed.

Growing unrest in the Depression years refocused the decades-old calls for government-funded advertising. The CFA increasingly saw it as an alternative to more controversial direct-aid programs. In October 1933, Joseph Paulhus editorialized in the *Canadian Fisherman* that the association was publicly opposed to unemployment insurance and believed that increased consumption was a better solution to the economic turmoil of the nation. According to Paulhus, it would take an increase in consumption of only five pounds per capita per year (a staggering goal of a 25 per cent increase in consumption during an economic depression) to propel the industry into "a period of unprecedented prosperity." The problem was clearly one of low consumption, since "there is never a question of not being able to supply the demand for fish."[90] The CFA then called upon the Department of Fisheries to provide $100,000 in advertising funding for "the benefit of the fishermen and fish trade in general throughout the Dominion." Such funding, the CFA insisted, would reduce relief payments to fishermen and "prevent the dissolution of this industry, which is one the great natural resources of the country."[91]

Perhaps more important than this public campaign was the return to leadership of the Liberal Party under Mackenzie King in 1935. During the election, members of Parliament began to react to the Price Spread Commission's spotlight on poverty and responded to the CFA's spin that such poverty could be resolved via increased domestic consumption. MP William Duff, for example, sent a telegram to the CFA promising to place before Parliament a bill supporting the cooperative's call for $100,000 in advertising assistance.[92] Much of this campaigning was, of course, covered extensively in the Maritimes press, and much of the political rhetoric focused on the need for government action to address poverty by stimulating the consumer economy.[93] Surprisingly, however, even as the CFA used the imagery of impoverished fishermen, the United Maritime Fishermen vocalized its opposition to just about everything the CFA said in its claim to represent the industry. William Groom of the UMF declared his opposition to the government-funded advertising campaign and to "many of the resolutions of the C.F.A., which represents the capitalist in the fishing industry," insisting that "advertising a high-priced food will not sell it." Instead, Groom supported the solution proposed by the UMF since 1927: banning trawlers.[94]

With the Liberal victory in 1935, the CFA got an industry-friendly minister of fisheries in Joseph-Enoil Michaud, of Restigouche-Madawaska, New Brunswick. The CFA began aggressive lobbying directed at Michaud's office. This activity did not go unnoticed. Liberal MP John James Kinley, of Queens-Lunenburg, Nova Scotia, perhaps the heart and soul of the Maritimes fisheries industry, wrote to William Found, who retained his position as deputy minister of fisheries after the transition to the Liberal government, that, "the Canadian Fisheries Association have a big voice in the fisheries affairs," and rightly so, due to their economic clout. Yet Kinsley also warned Found that the government had a responsibility to look after the common fishermen, because "the real producer, the fisherman, has no unity and is unable to enforce his demands because he has no bargaining power." Kinsley did not mention fishermen's desire to ban trawlers, and Found provided only a vague response to his note. Other than Kinsley's one memo, there is little record of the Department of Fisheries, or any other MP except Kinsley, spending much time listening to fishermen's concerns. The power of the fishermen seemed as weak in the halls of government as at the auction of their catches.[95]

Judging from the public discourse in the newspapers in 1935–36, it would appear that everyone was concerned for the well-being of fishermen across Canada, but especially of those in the Maritime provinces.[96] Yet judging from the internal communications in the files of the Department of Fisheries, which include extensive communications with the CFA and virtually none with fishermen's unions, such rhetoric might be accused of being empty political discourse. Much of Michaud and Found's effort in 1935 and 1936 was to secure funding for an advertising campaign to increase domestic consumption of fish, even though the rhetoric used to justify spending this money to buttress of private-sector economy relied largely upon public concern for the fishermen's poverty.

Michaud represented a sea change for the idea of government-funded marketing to aid in the recovery, stability, and growth of a private-sector industry. Writing to one of the industry leaders, R.W. Widdess, president of Seaport Crown Fish Company in Vancouver, Michaud echoed the very words of the Canadian Fisheries Association: "There can be no room for doubt that if reasonably adequate markets were available most of the difficulties under which our fishing industry has been, and continues to labour would disappear."[97] Frederick Wallace, the new president of the CFA and former editor of its *Canadian*

Fisherman, lobbied Michaud in December 1935 with an eleven-page memorandum. In it, Wallace outlined the history of economic stagnation facing the fisheries industry since 1884. He also outlined the various steps taken by the federal government to promote the industry's better economic health, including subsidizing railroad freight for fresh fish for eight years beginning in 1909. Since 1918, Wallace concluded, the fisheries industry had been trapped in a debilitating boom-bust cycle that prevented the necessary economic stability for long-term investment strategies. The $34 million the industry had netted in 1934 was the same amount it had reached in 1911, but costs of operation had since doubled. Wallace believed that deepening the industry-government partnership would "inaugurate a period of safe, sane, progressive and well-planned development of the vast fisheries which is the heritage of our country."[98]

Such progress rested on consumption: "The crying need of the industry today is markets," Wallace insisted. "Production presents no problem." The industry had the workers and technology to produce twice as much product if markets could be secured. With proper encouragement, he argued, the Canadian market could consume 100 million pounds per year, and past campaigns had been successful in relation to the modest funds provided. As with previous claims of success, Wallace provided no evidence. He reiterated the time-worn rhetorical defence of this demand for government-funded advertising: it was the only way to help impoverished workers across the industrial landscape. Fishermen were "the principal sufferers" and had not been able to increase or diversify their income, remaining wholly dependent upon the industry's well-being. It was the fishermen and the shore workers "who [stood] to benefit most by an increased market for their products."[99]

Various industry groups and municipal boards of trades in cities with vested interests in the fisheries sent supporting documentation to Minister Michaud. In one such letter, the Halifax Board of Trade took a tangled approach that argued for limited government involvement – "the future of the Industry does not lie along the road of marketing acts, inspection or government regulations, but rather upon the free and unhampered enterprise of those vitally interested in the Industry" – while also calling upon the government to provide direct aid to the industry for "necessary assistance" that would sow the seeds of future industry prosperity.[100] The Board of Trade seemed untroubled by calling for limited government regulation while at the same time asking for more government aid.

With the weight of the industry fully behind him, Michaud went to the Privy Council and successfully lobbied for $200,000 in appropriations "to aid in expanding the sale of the products of Canadian fishermen in foreign and domestic markets."[101] Slightly more than half of this money, $117,000, would be spent on domestic advertising and marketing.[102] The Department of Fisheries then hired E.W. Reynolds and Company of Toronto to handle all advertising west of the Maritime provinces and the Atlantic Advertising Agency of Sackville, New Brunswick, to cover the Maritimes. And so, after nearly a decade of lobbying, the CFA finally got the government funding it believed was essential to stimulate domestic consumption of fish and spark economic recovery for an industry long in depression.

For the next several years, the Department of Fisheries cooperated with the CFA in a large nationwide advertising and marketing campaign to increase domestic consumption of fish and fish products. The campaign used three lines of discourse. The public discourse – the content of the advertisements and media marketing directed to the consumers – primarily focused on the health and economic advantages consumers would receive from switching more meals to fish. The political discourse, used by industry leaders in the trade press and by government politicians and bureaucrats in the halls of government, emphasized the importance of increased domestic consumption as a direct means of improving the well-being of the producers, both at sea and on shore. Finally, the internal communications between the industry and the Department of Fisheries largely focused on industry-wide recovery, stability, and growth. The first of these three has been discussed in chapters 1 and 2 on health and family well-being. It is the balancing act between the latter two, public discourse and internal communications, that needs to be more fully deconstructed in this chapter.

George Akins, of E.W. Reynolds, and H.F.S. Paisley, the new director of publicity for the Department of Fisheries, together oversaw most of the campaign. Akins spent the department's money on advertising in major national magazines like *Chatelaine, Canadian Home Journal, MacLean's, La Revue Populaire*, and the *National Home Monthly*, as well as in twenty-four major metropolitan dailies, fifty-three provincial dailies, twelve farm papers, and numerous Canadian weeklies, grocer and retail trade journals, and ethnic newspapers (figures 4.1 and 4.2). Much of the advertising content has been addressed in previous chapters; the justification of the campaign, however, was more directly

Fig. 4.1   "My Dear ...," Any Day a Fish Day campaign advertisement, *Canadian Grocer*, 15 October 1939, 29. The Department of Fisheries needed not only to convince more Canadians to eat fish but to convince more retailers to stock fish for consumers. Advertisements placed in retailers' journals like the *Canadian Grocer* claimed unmet demand and urged retailers to answer the call.

addressed in editorials secured by Akins and Paisley in industry-friendly newspapers and magazines, and in internal communications between the Department of Fisheries, the CFA, and E.W. Reynolds.

The campaign was initially to cover the winter months of 1936–37, the peak season for fish retail sales in Canada. It would be followed by three additional campaigns that, in combination, lasted until 1940, again focused on the winter months. Although the content changed slightly for the final two campaigns in 1938–39 and 1939–40, the marketing methods and basic justification remained relatively consistent. In 1936, E.W. Reynolds identified two key justifications for the campaign. The first was to "promote national health" by informing Canadian consumers of the health value of eating more fish. This was set in the context of "national health" based on a growing belief that a nutritional crisis existed in Canada and that many Canadians were "starving in the midst of plenty" (see previous chapter). The vast majority of the content of the government advertising thus rested on

SERVE

fish

FOR FAMILY HEALTH

Watch that boy of yours grow when you give him plenty of good Canadian fish at mealtimes.

He'll grow sturdy and strong, as healthy as a horse, and that appetite of his will relish the delicious change that Fish makes in his diet. Fish is the greatest known food source for Vitamin "D", the sunshine vitamin so essential for growing children.

The whole family will enjoy Fish. It can be served in so many appetizing ways . . . there are over 60 varieties of Canadian Fish and Shellfish from which to choose . . . and no

other food gives you so much health-building nourishment for so little money.

No matter in what form you like Fish, it is available to you in Canada from Coast to Coast. Serve it often . . . it makes a welcome change in your family menus.

DEPARTMENT OF FISHERIES.
OTTAWA.

Ladies!

WRITE FOR FREE RECIPE BOOKLET

Department of Fisheries, Ottawa
Please send me your free 52-page Booklet, "Any Day a Fish Day", containing 100 delightful and economical Fish Recipes. 71

Name

Address

ANY DAY A FISH DAY

Fig. 4.2 "Serve Fish for Family Health," newspaper advertising clipping, 1937. The Department of Fisheries invited "Ladies!" to "Write for a Free Recipe Booklet" to help fulfill their obligations to feed their families healthy meals. The marketing linked improved individual health to the collective well-being of a nation suffering from what many believed was a nutritional crisis.

the nutritional value of fish and seafood products. The second justification, less overt in the public campaign but central to the political
and industry rhetorical justification for spending public money, was
the need to "stimulate [the] Canadian Fish Industry." Increased domestic consumption of fish would, according to the press copy, provide
the industry with "a stimulus that would benefit not only the Industry but the Dominion at large."[103]

Meanwhile, the Atlantic Advertising Agency, because it focused
exclusively on marketing fish in the Maritime provinces (New
Brunswick, Nova Scotia, Prince Edward Island, the Magdalen Islands,
Gaspé, and other Atlantic coastal communities of Quebec), could
appeal directly to the producers and their families and neighbours. In
one marketing spot, the firm called upon Atlantic Canadians to "Put
Down Your Bucket Where You Are!" After retelling the common folk-
tale about a ship thought to be lost at sea without fresh water, the
press release asked teachers to urge boys to become fishermen while
also calling upon the community to support their labour. "If Canadian fish and shellfish were served regularly, as often as three times a
week in every home in the Dominion, the increase in employment
and in general national prosperity would be truly amazing." The
agency convinced the Department of Education in both Prince
Edward Island and Nova Scotia to forward the message in official
communications to principals and teachers within the provinces.[104]
In a 1936 radio spot, the agency urged, "School children should
become active participators in this campaign, by being taught that
their consumption of fish contributes to one of Canada's basic industries."[105] Another radio spot highlighted the value of national heritage
and the needs of labourers: "By increasing home consumption of
these products we shall be learning to appreciate our good heritage,
and shall be helping the fishermen to whom the fisheries mean bread
and butter."[106] However, these direct appeals to national, regional, or
industrial economic recovery, although central to the political rationale for the campaign, were rare in the advertising that overwhelmingly focused on individual health and family economic needs.

References to the national importance of the fishing industry
appeared in editorials and news coverage of the campaign and the
Department of Fisheries more often than in the direct consumer-
focused advertising. This marketing, which the advertising firms
insisted was separate from the advertising campaign, remained central
in the department's public relations efforts. Editors across the nation

eagerly lent support. The *Charlottetown Guardian*, for example, proclaimed in a February 1937 article, "The Hon. J.E. Michaud is the first Minister of Fisheries fully to realize the necessity of developing the industry in the only practical way – increased consumption." The newspaper shared the view of both the CFA and the minister of fisheries that "advertising and consumption go together, and if domestic markets are to be catered for, intensive publicity is necessary to 'sell' the goods to the prospective consumer."[107]

Yet it was not enough to just articulate that the campaign would aid the industry. Within the context of the Depression, more specific references to the necessity of improving impoverished fishermen's lot were central. In 1938, after two year of aggressive advertising, the CFA's *Canadian Fisherman* claimed, without evidence, that the Canadian public had become more "fish conscious," and that consciousness included a concern for the well-being of the producers within the industry: "Fishermen in the distressed areas of the eastern seaboard would benefit greatly" when consumers used their purchasing power to promote societal improvement.[108] *The Ottawa Morning Journal* similarly highlighted the social justification of government spending to advertise products for a private, for-profit industry because of its positive impact on poverty and underemployment, noting that such efforts would "improve the condition of the thousands of families dependent on this industry for their livelihood" and be of "benefit not only to these primary producers but to the consuming public," bringing economic prosperity to the entire nation.[109] In a 1939 summary of the three-year-long campaign, E.W. Reynolds maintained that "in the year 1935 the fishing industry of Canada was in the throes of one of the most depressed conditions in its history" and "the Canadian fisherman was beset with innumerable difficulties in the marketing of his product." The federal government launched its campaign for the "purpose [of] the educating of the consumer of the qualities of fish as a health food,"[110] and, according the firm hired to conduct it, had brought industrial recovery and improved incomes for impoverished workers and had a positive impact on national health and well-being.

Fisheries corporations, of course, largely celebrated the campaign. C.J. Murrow of Lunenburg Sea Products, for example, wrote to Found in November of 1936, "I think the work your Department and the advertising agency employed by you is doing real good and I have heard several favorable comments."[111] But was it indeed successful?

Did it increase fish consumption or raise the "fish consciousness" of the Canadian public? Did it have a positive impact on the industry as a whole? Did industry stability and growth trickle down to the common labourer? Hard data on increased domestic consumption of fish was hard to come by. The industry trade paper *Maritime Merchant* claimed that per capita consumption had increased from twenty-one to thirty pounds (a staggering increase of nearly 50 per cent) but gave no source for the claim.[112] *Marketing*, an advertising trade journal out of Toronto, insisted there was a 25 per cent increase in fish sales as a direct result of the advertising and that per capita consumption had increased from twenty-one to twenty-six pounds, "according to authorities" – neither named or cited.[113] Trade journals like the *Maritime Merchant* and *Marketing*, of course, had vested interests in claiming that such advertising was successful. A 1939 report from E.W. Reynolds stated that consumption of fish in Canada had increased by 35 per cent (not 50 or 25 per cent), and that per capita consumption had risen from eighteen pounds to twenty-nine.[114] Again, the company provided no references for those figures. The differences in per capita annual consumption figures suggest that no real evidence of the campaign's impact existed.

This vagueness persisted through 1940. George Akins insisted that "although exact figures are not available, there can be no doubt of the fact that the increase in consumption of Canadian fish products has been due in the greatest possible measure to the powerful campaigns of publicity and advertising continuously carried on during the past four years by the Department of Fisheries."[115] The annual reports from E.W. Reynolds to Michaud seemed to focus mainly on the number of advertisements placed, their distribution across different papers in different regions, and even the number of lines such ads consisted of, rather than assessing the impact on sales.[116]

Yet even without exact figures for increased sales, many interested parties argued that the campaign was having a positive effect. Allan McLean, of Connors Brothers in Saint John, believed that the campaign's success, rather than in immediate measurable sales increases, lay in public awareness that would have long-term positive benefits for the industry. He wrote to the CFA, "It has been a great educational campaign in acquainting the people of Central Canada with more knowledge about the different varieties of fish that we have on both the Pacific and Atlantic Coast, and also in informing them of new ways to prepare sea food."[117] Michaud's main measurement of success was the number of requests his department received for the free fish cookbooks, which he reported as being between three hundred and six hundred a day. In 1938, William Found stated that the department had sent out nearly 200,000 cookbooks.[118] The *Saint John Telegraph Journal* took this as "striking evidence" of success, while also noting "figures from industry" that proved increased consumption – again giving no specific figures. The newspaper also reported the statement of C.J. Murrow, president of the CFA, that "domestic consumption has increased greatly and the position of fishermen is better than it has been for some time."[119] At the cooperative's annual meeting in Toronto, CFA secretary R.W. Gould supported Murrow's claim: "There is not the slightest doubt in our minds but that the advertising had a salutary effect upon increasing the consumption of fish." No doubt they believed it.[120]

Despite these claims, the same industry voices also worried about a possible glutted market in 1937. After a year of aggressive advertising, the industry still produced more product than the domestic market could consume. In a 26 April 1937 resolution by the directors of the CFA, these industry leaders, who had just ten days before told the press that the consumer demand had "increased greatly" to the benefit of the fishermen, now warned the Department of Fisheries of "a grave danger of a glut on the market." The situation could only be addressed by "the developing and widening of the domestic markets for fish, in order that the present production be properly absorbed at a price commensurate with the cost of production."[121] The contradiction between greatly increased consumer demand and a dangerously glutted market did not seem to bother Michaud, who was in total agreement with the CFA that the advertising work had to be continued. Speaking before the association at its annual meeting in Toronto in mid-September, he proclaimed that fish exports had increased by 25 per cent and "domestic fish business had shown a great increase."[122] Less than a month later, he told a meeting of the Privy Council that the campaign had "beneficial results to the fishing industry."[123] He subsequently secured an additional $100,000 for the 1937–38 advertising season, half of what had been provided in 1936–37.[124]

The 1937–38 campaign continued to emphasize the value that fish provided to the consumer in terms of health and affordability. In 1938, the Canadian Fisheries Association again petitioned Michaud for marketing funding for the 1938–39 season. Telegrams from CFA members flooded Michaud's office with thanks for his hard work, and Gould, the association's secretary, insisted (without evidence) that the campaign had so elevated domestic consumption that "the unemployment situation so far as fishermen in Canada are concerned has been materially relieved."[125] In a subsequent letter, Gould stated that from his interviews with wholesalers and retailers, he estimated that domestic consumption was up by 15 to 20 per cent. In spite of this, he admitted, "The available statistics do not show a true picture of the very beneficial effects of the Campaign so far as actual comparisons were concerned."[126] Instead, he insisted that the mild winter of 1937–38 would have seriously depressed fish sales, as seafood did not sell well in warmer weather, had it not been for the government's advertising. While there was no hard data to support any of these claims, and some metrics showed no growth at all, the fact that fish consumption did not significantly decrease was evidence enough of success. All these statements were extremely influential. In his reports to the Privy Council, Michaud regularly reported that the CFA, a "representative of important branches of the industry in all parts of the Dominion," insisted that the campaigns were successful, and therefore they must be.

The political rhetoric continued to revolve mainly around impoverished fishermen. For Michaud, this represented a clear change in policy within the Department of Fisheries, which, according to him, had to look beyond fish in determining what steps it would take. In an opening address to the 1938 Sea-Fisheries Conference in New Brunswick, Michaud stated, "The problem of the fisheries is no longer merely a fish problem – it has become a fishermen's problem, it has become a social problem – and this is why it is much broader than possibly was intended by those who framed the constitution under which we have to work and labor." It was time for the department to move beyond conservation, inspection, and regulation and embrace the new needs of social welfare and protect the ability of fishermen to "make a livelihood and to derive some compensation for their labor in order to take care of their families … to provide for themselves." One might think that Michaud would then turn to such themes as minimum prices paid to fishermen or fishermen cooperatives, but instead he continued to insist that increased domestic consumption was the key solution to the fisheries problems and that the advertising campaign had resulted in a 35 per cent increase in domestic consumption.[127] Michaud's rhetoric was largely successful in the halls of government. Stating in his 1938 presentation to the Privy

Council that "assistance is still needed by the fishermen in expanding the demand for their catches so that they may obtain an adequate return from their labours," he succeeded in getting an additional $130,000 approved for the 1938–39 season.[128]

The industry was quick to celebrate. Gould wrote to Michaud in April 1939 insisting that the industry would not have survived the worst years of the Depression without the advertising support from the federal government, and that the campaign "had a most salutary effect upon the economic condition of the actual fishermen of our coast, one of the prime purposes for which the campaign, we believe, was designed, and in which direction the expenditure of the grant, in our opinion, amply justified itself."[129] Within the context of the Depression, and with the findings of the Price Spread Commission of 1934 still hanging overhead, it is not surprising that successful political rhetoric referred to improving the living conditions of those catching the fish. Yet in all of these references, no data were presented to provide clear evidence that the income and well-being of the fishermen and their families and communities had improved; nor was any evidence presented that any improvements were directly related to increased sales volume from more aggressive advertising. Nevertheless, the assumed success of the 1938–39 campaign led to additional funding. Another $200,000 was provided for the 1939–40 season. This time nearly all of it, $180,000, was to be spent on the domestic market, making it the best-funded campaign of the 1936–40 period.[130]

When Canada went to war in September 1939, the government, and the Minister of Fisheries, had to consider whether such expenditures were appropriate, given the new needs of the nation. At first Michaud believed the existing appropriations for the 1939–40 season would continue as planned. George Akins prepared a press release for circulation in as many newspapers and magazines as possible, articulating that the department would continue the campaign in the "interest of general business stability and to contribute to the maintenance of public confidence in a time of emergency."[131] Akins and Paisley quickly changed some ad content to emphasize war preparedness, but, in general, the 1939–40 campaign remained similar to the previous three ones.

The war provided unique challenges to specific sectors of the industry. Canned lobster appears to have been particularly hard hit; in 1939, the United Kingdom placed canned lobster on its list of luxury items and prohibited its importation.[132] Minister Michaud responded immediately that the best way to aid the lobster industry was to launch a massive advertising campaign. The Canadian federal government purchased 55,000 cases of canned lobster, relabelled them "Canada Brand Lobster," and launched an energetic effort to sell them.[133] Like previous campaigns, this one utilized popular national magazines, major metropolitan weeklies, smaller regional and city dailies, and radio to market canned lobster as a healthy, safe food.[134] Yet, unlike the earlier campaigns, this one was more overt about the need for Canadian consumers to think about the societal implications of their purchases (figure 4.3)[135] Many advertisements told the consumer the industry was "facing a serious emergency." The words "Be Patriotic! Serve Lobster – It's a Pleasant Way to Serve Canada" were paired with an image of a lobster proudly waving the Canadian ensign.[136] The language of war also entered the lobster campaigns. Monica Mugan's radio talk, for example, approved by the Department of Fisheries, described lobsters as "autocratic, bullying creatures who think nothing of tearing each other to shreds," and who "get into drunken quarrels and rip at each other … and others go off into corners and have laughing jags." Mugan's rhetoric conveyed that eating lobster taught brutish enemies of civilization an important lesson as it provided economic stimulus for Canada.[137]

Most industry and industry-friendly papers celebrated the success of the canned-lobster campaign, which continued into 1941 with an additional $50,000 in government funding.[138] In this case, there appears to be some evidence of increased domestic consumption, although it was largely the result of intense government control of the market. Not only did the federal government own a large stock of the product but it also set up a Lobster Control Board to regulate price and volume at all levels of production and sale. The government was able to do this because it listed lobster sales as necessary under the War Measures Act. Moreover, the Department of Fisheries was able to convince the Canadian military to add canned lobster to camp rations, selling off a very large percentage of government stock to the government itself.[139] Although hardly an example of free-market enterprise, the strategy resulted in clearing out all existing overstock and increasing the prices paid to lobstermen in Atlantic Canada.[140] It was just the kind of "success" that industry leaders could point to in encouraging further government aid in marketing. Yet it is important to note that the success was not just, or even mainly, a result of increased advertising.

Fig. 4.3   "Women of Canada – You Can Help!," advertisement, *MacLean's* magazine, 15 August 1940. When the Second World War erupted in 1939 and Great Britain placed Canadian lobster on the luxury list, the Department of Fisheries bought up large quantities of Canadian lobster, rebranding it as "Canada Lobster" (a government-owned enterprise), and dumping it onto the domestic market. Just as housewives were called upon to fix the fisheries' economic crisis and the national health crisis by purchasing more fish, they were now asked to contribute to the war effort by rescuing the lobster industry.

The war years brought an end to the advertising campaign for four main reasons, the first two relating specifically to the war. First was the growing realization that spending money to encourage Canadians to eat fish in a time of military crisis was bad public relations. Second, as during the First World War, Europe's food needs increased all food production in North America as well as foreign exports, and so advertising was unnecessary. Third, some industry and government leaders believed the campaigns of 1936–40 were less successful than claimed. Finally, many believed the advertising had failed to increase consumption because it did not address the core problem for the Atlantic fisheries: poor and inefficient productivity that resulted in a low-quality product.

Fishermen's unions had long voiced opposition to the campaigns they saw as only aiding large trawler operations and their poor-quality fish.[141] But it was not only fishermen who saw a fallacy in an advertising campaign designed to stimulate industrial recovery and prosperity for producers. John James Kinley, MP for the Lunenburg district of Nova Scotia, noted succinctly in 1936 that "nobody will eat fish to help the fisheries or fishermen. They will eat fish when it is in their own interest."[142] What was shocking was the sudden change of perspective among industry leaders. In June 1940, F.T. James, the operator of Toronto's largest fish wholesaling and retailing business, actually reported that there had been a decline in total volume of sales during the campaigns. George Akins, the campaign's architect, tried to put a positive spin on that information, arguing that the decline in volume was matched by an increase in value, which resulted in a general stability, or even increase, in industry profits and fishermen's wages. He concluded, "As our main endeavour was to increase the remuneration to the fishermen of Canada, the writer feels the main purpose of the campaign for 1939–40 was fulfilled in this instance."[143] Yet earlier communications had clearly articulated that the goal was to increase volume, not value, of product sold.

Perhaps most shocking was the statement of the former president of the CFA, C.J. Murrow, president of Lunenburg Sea Products and once one of the campaign's most vocal supporters. In June 1940, however, he wrote to Donovan Finn, the new deputy minister of fisheries, "Frankly the results of these campaigns have not been as good as I had hoped they would be." While Murrow felt they had improved consumer knowledge, they had not stimulated increased consumption. It would therefore be unjustifiable for the government to continue spending public money on the campaign when "every dollar is needed for war effort." As wartime demand for food in Europe would no doubt increase demand for fish, marketing would be unnecessary. He concluded, "In any event we hope the Government will cut to the bone any expenditures of this kind during the war."[144]

By the war's end, critics were arguing that the real problem with the Atlantic fisheries was inefficiency and poor product. In 1943, Nova Scotia's provincial government hired Stuart Bates, a commerce professor at Dalhousie University, to write a comprehensive report on the state of affairs within the province's fishing industry. While Bates viewed the Pacific fisheries as a modern operation exploiting the resources to its maximum capacity, he saw the Atlantic fisheries as "showing all the characteristics of small-scale competitiveness – the lack of organization among fishermen, the jealous individualism among the fresh fish firms ... and among the salt fish exports."[145] The failure of the Atlantic fisheries he saw as rooted in the industry's failure to adapt to new realities of the food industry prior to the war. When most other food sectors modernized via cooperative movements and partnerships, inspection and quality control, integration of production, distribution, and retailing, as well as crafting brand names and attractive packaging and advertising, the Atlantic fisheries remained dependent upon nineteenth-century food products like salt fish. When consumers demanded fresh food, the fisheries were unable to respond.

Bates's emphasis on quality, however, was not the same as that expressed by fishermen unions in their broader campaign against trawlers. On the contrary, Bates celebrated trawlers for opening the fresh fish market with efficient and reliable means of landing fresh fish. He concluded that 10 per cent of fishing labour landed over 60 per cent of the catch, and "the more highly capitalized is the catching equipment, the greater is the mobility and catching power, and the more constant the fishing operations." He bemoaned that the "trawler controversy" had resulted in government policy designed to keep as many men employed as possible rather than "strengthening the competitive power of the fishery against the growing modernization of the other food industries with which Maritime fish had to compete." The result was that, by 1944 in the Maritime provinces, fishing "depended on men," whereas in New England and on the Pacific Coast fishing depending on technology.[146]

For Bates and others, postwar development of the fisheries had to take into account consumer demand for high-quality, reliable products. The advertising campaigns of 1936–40 failed because they tried to push a low-quality product, just as William Fraser had forecast in his 1934 report for the Price Spread Commission. Increased consumption of fish in postwar Canada would require intense modernization of the industry and increased quality-control measures by both industry and government. The Atlantic fisheries had made virtually no progress in this endeavour throughout the 1920s or '30s, and even during the wartime boom there was little investment in modernization or quality control. Bates predicted that, once wartime demands subsided, the fractured nature of the industry would result in continued long-term decline. His ideas deeply influenced the federal Department of Fisheries, which immediately after the war completely re-evaluated its perspective of the consumer market by embracing improved production as a prerequisite to increased consumption.

The Department of Fisheries continued to support increased production, even if it meant limiting its traditional role as a monitor of conservation. It often tried to check public discourse that suggested fish-stock depletion. For example, CBC program director Richard S. Lambert produced a series of audio presentations on Canada's resource conservation for high-school audiences, its initial draft for the fisheries piece, "Forty Fathoms," warning, "It is a fallacy that the sea provides an inexhaustible supply of fish. Unless regulatory measures are taken, unlimited exploitation means that fishermen have to go further out, spend more time and effort to bring back a given catch."[147] H.F.S. Paisley wrote to Lambert, "It would be better if your references to the supply of fish as not inexhaustible were made a little less sweeping." The statement, "even though literally true, might [give] mistaken impressions to the uninitiated."[148] Paisley specifically asked Lambert to add words like "may" and "some" to add vagueness to the warning.

During the final years of the war, the department increasingly turned to the theme of increased efficiency in production. Paisley, for example, wrote to Colonel A.L. Barry,[149] chief supervisor of Fisheries in Halifax, emphasizing that the key to growth in the Maritime provinces was "production that is efficient." In very similar language to that of Bates, Paisley noted, "This is a food industry. It cannot hope for steady success unless it supplies consumers with products which are comparable to other foods in quality." His words were a clear endorsement of the trawler operators' claim that they provided the market with regular, reliable, and uniform fish products.[150]

The Canadian fisheries industry, like other natural-resource-based economies, had collapsed in the 1920s. For these industries, the Depression began around 1920 and did not ease until wartime demand sparked recovery in the 1940s. Thus, fishermen, business leaders, and government bureaucrats struggled for two decades to find a solution to what appeared to be perpetual depression and poverty in fisheries sectors. For the Canadian Fisheries Association, the industry's leading voice, the answer to the problem was clear and consistent: government investment in a campaign to increase domestic consumption of seafood. Constitutional frameworks like the British North America Act limited the ability of the federal government to address poverty or economic recessions and depressions and so to aid a natural resource industry unable to control either production or consumption. By the mid-1920s, having made very little progress in rationalizing or understanding the variables of supply, the fisheries industry turned to the variables of demand.

The moment was now right for a new perspective on industry recovery, stability, and growth. The age of consumerism was well established, and few business leaders in Canada doubted the power of consumption as an economic driving force. From sporadic beginnings in 1924 and 1925, and through more consistent endeavour between 1934 and 1939, the CFA and the Department of Fisheries conducted a series of advertising campaigns to increase domestic consumption of seafood to address the seemingly intractable economic dilemma of the perpetual economic decline of the fisheries industry and fisheries labourers. In this sporadic, roughly twenty-year campaign, the department and the fisheries association remained surprisingly committed to recurring themes easily recognizable to any observer of contemporary seafood advertising. The campaigns of the 1930s and '40s attempted to capture the new consumer obsession with food health by marketing seafood as a wholesome addition to the family diet. Within the context of the Depression, the campaigns also emphasized the thrift and cost-saving advantages of eating more fish. Finally, they captured the heightened sense of Canadian nationalism, sparked by the Statute of Westminster in 1931 and carried throughout the war years, emphasizing the patriotic value of buying Canadian-produced seafood and using one's consumer power to buttress the well-being of fellow Canadian producers.

Yet the end result was disappointing. Seafood consumption remained relatively low in Canada, even declining after the Second World War, and the nation's seafood industry continued to rely on an export market over which it had nearly no control. Despite earlier claims that per-capita consumption had increased to as high as thirty-nine pounds per year, when the Department of Fisheries finally developed a matrix for measuring per capita consumption in 1952, it found that in 1939 per capita consumption was probably around twelve pounds and had increased to only fourteen pounds by 1951.[151] The persistent insistence throughout the 1920s, '30s, and '40s that the solutions to economic downturns lay in increased consumption was badly flawed.

5

Fish Will Win the War:
Patriotism and Seafood Rationing
in the Second World War

The period between 1939 and 1952 marked a new chapter in the history of fisheries consumerism in Canada. Wartime fisheries policy mirrored that of other food production, distribution, and consumption. The Department of Fisheries encouraged Canadians to increase their consumption of fish to preserve other food resources for export to war-torn Europe, to build healthier bodies in the name of war readiness, and to use their consumer dollars to buttress an industry central to national economic viability. Following the war, the main debate was between the competing ideas of directly aiding Europeans with food supplies or assisting them in rebuilding their capacity to produce their own food. In terms of the fisheries, most of the leadership in organizations like the United Nations Relief and Rehabilitation Association (UNRRA), the European Recovery Program (also known as the Marshall Plan), and the UN Food and Agriculture Organization (FAO) wanted to help Europe develop self-sufficiency in food production, including fisheries. The Canadian Department of Fisheries, however, noted that Europe had not been self-sufficient in fisheries production since the discovery of the fisheries resources of North America in the sixteenth century and that it was ahistorical and inefficient for Europe to be self-sufficient in fisheries. More pressing, however, was the realization that a self-sufficient European fishery would mean a massive loss in customers and revenue for Canadian fisheries. In the interim, many of these same organizations purchased massive quantities of Canadian seafood for European food-aid programs. The immediate result of both wartime and postwar food policies was the Canadian fisheries industry's increasing reliance upon government purchases. These guaranteed markets encouraged production well beyond what natural markets would have demanded. Once the government-purchase programs ended in the early 1950s, the Canadian fisheries industry once again faced glutted markets caused by overproduction in many key sectors, especially in non-perishable fish products like canned seafood. Once again, short-term analysis of consumer demand (this time buttressed by massive government purchases) resulted in serious long-term mismanagement.

The wartime pro-fish propaganda of the Second World War released by the Department of Fisheries and its industry partners focused on three main messages.[1] First, similar to the campaigns during the First World War, much of the pro-fish marketing deployed by the Department of Fisheries and the Canadian Fisheries Association (CFA) focused on eating more fish domestically to conserve meat for use in Europe. Second, Canadians were urged to eat more fish for their health. Although much like the messages discussed in chapter 2, within the new context of global war, Canadians were told that their individual health and well-being were paramount for the nation's success in defending liberty. Third, the propaganda argued that consumption of domestically produced seafood would promote the national economy and aid in Allied victory. All three messages played equal roles during the war.

Again, as during the First World War, the Department of Fisheries saw opportunity in wartime rationing of meats as a means to promote fish consumption along patriotic lines. According to Joseph-Enoil Michaud, minister of fisheries, "increasing the consumption of Canadian fish becomes, in fact, a matter of national importance under present conditions." He called upon Canadians to restrict their consumption of "certain classes of foods" so those products could be sent to Great Britain and France, and to instead increase their consumption of seafood as a matter of national and global security.[2] In summer 1941, the government asked Canadians to reduce the amount of bacon they ate for breakfast so it could export it to the war-inflicted people of Great Britain. The fisheries industry and the Department of Fisheries saw this as an opportunity to restore a traditional fish-based breakfast and blasted newspapers with pro-fish breakfast propaganda. By eating fish every morning, "patriotic citizens" could directly aid in the ability of their British allies to withstand the trauma of war.[3]

Canadians were also told that eating more seafood would contribute to national health, which was essential for victory. By the time
Canada went to war in September 1939, there was already a well-
entrenched concern for national health based on the supposed hidden hunger crisis of the 1930s and a general fear of malnutrition. This
fear was augmented by high rates of rejection of armed forces volunteers due to poor health and strength. Canadians were thus often told
that "Health Is One of Our Resources" and that it was everyone's patriotic duty to maintain their health and strength for war-related activities, whether for battle or wartime industrial production. "In the
mighty mustering of Canada's resources to aid in the defense of Liberty," one propaganda piece from the Department of Fisheries proclaimed, "the physical fitness of her people is imperative." Canadians
could find "no better source of health and strength than Canadian
Fish and Shellfish" for the courage, determination, energy, and alertness needed to win the war. The protein and vitamins found in fish
would "build vigorous health and alter fitness" for both factory workers and soldiers. The "keep fit and do your bit" message appeared
across the food industry, and the fisheries industry and its government allies were not shy about piling on that rhetorical trope. Yet it
was equally important to "insist on Canadian Fish" to "help strength-
en the national economy in these difficult days."[4]

The promotion of the national economy during a time of global
crisis constituted the third rhetorical device used in this propaganda.
Active consumerism promoted strong national industry, which was
central to national strength during the war. The fisheries remained
one of the most important national economies. In a message circulated
to the nation's newspapers in September 1939, Minister Michaud
noted that the nation's fisheries had been important to the national
economy during the pre-war years, but in the midst of a global war
for liberty, the economic well-being of the fisheries industry was a
"vital necessity."[5] Consumers were told to "be patriotic" and purchase
food vital to maintaining national economic strength.[6] In July 1941,
the *Guelph Daily Mercury* called upon readers to "do their part" and
promote the economic well-being of an important Canadian industry
and the labourers who worked for it. Since fish was healthful and
inexpensive, "the housekeeper desirous of falling in line with the
needs of her country will find this desire backed very sensibly in a
booklet compiled by the department," which provided helpful,
healthy, and easy-to-prepare fish dishes.[7]

As Ian Mosby notes in his 2014 book, *Food Will Win the War*, Canadians often saw the nation's vast food resources as a prime contribution to the Allied war effort. Mosby cites two Wartime Information Board polls in 1942 and 1944, in which 38 and 30 per cent of Canadians polled, respectively, cited food as the nation's greatest contribution to the war effort. Canada's wartime food production and distribution to Europe, and in particular Great Britain, brought what Mosby writes was an "unprecedented state intervention in Canadian food production." Yet Mosby accurately articulates that, of two avenues through which the Canadian federal government managed food during the war, production was perhaps less important than consumption, as Mosby notes; "The federal government intervened directly in the operation of the nation's kitchens on an unprecedented scale." Such intervention came in two forms: food rationing and nutritional control. The government called upon Canadians to conserve key food products by eating less or eating different commodities (like fish), and to improve their nutrition to be better prepared in mind and body for wartime effort both at home and at war. For Mosby, the government's control of both production and consumption was not just an intervention into food habits but part of the larger history of the development of the welfare state in Canada. The intervention in food paved the way for an active government following the war.[8]

During the war years, the influence of the new science of nutrition on food selection rapidly built upon its pre-war popularity, thanks in part to active government support for nutritional research and propaganda. The fear of malnourishment was no longer entirely a question of equity and social justice; in the context of war it became an issue of national security. Leading nutritionists in Canada, Great Britain, and the United States agreed that to achieve full bodily potential, individuals should consume more than the minimum nutritional requirements. James Orr in Great Britain and Hazel Stiebeling in the United States led the charge for maximum nutrition. In Canada, Leonard Marsh, Harry Cassidy, E.W. McHenry, Margaret Bell, and Lionel Pett communicated these ideals to the population. The claims seemed proven during the initial days of the war when Canada's men appeared to be insufficiently healthy for combat; 43 per cent of Canada's first fifty thousand volunteers were initially rejected for medical reasons. Although the rejections resulted from a range of medical issues, nutritionists pointed to the impact of sub-optimal nutrition and recommended supervised diets to better prepare recruits for military service.[9]

This perceived health crisis also affected war industries. The economic necessities of production were omnipresent within the nutritional discourse of the war. Whereas before the war nutritionists were more apt to discuss deficiency diseases and equity, during the war they spoke of reduced productive capacity. Mosby delves deeply into the "manpower" crisis facing key industries. Historians have long studied the impact of wartime labour shortages with a particular focus on finding replacement workers; nutritionists focused on increasing the physical capabilities of those male workers who remained in war production plants. Improved nutrition would enable workers to work longer and harder shifts, reduce days missed due to illness and fatigue, and improve mental alertness to reduce costly errors. Better nutrition would solve the crisis in masculine strength and vigour that sapped Canada's war efforts. Mosby concludes that these ideals, although given new power within the context of global war, were rooted in prewar social constructs of reform and the Protestant masculine ideal. In this context, he illustrates that nutritionists' championing of better food health made a distinct shift during the war away from questions of equity, income, and access towards individual informed choices. Seeming to re-embrace one angle of the elite-driven Progressivism of the early twentieth century, the wartime food campaigns articulated individual responsibility more than societal reform.[10]

Nutrition thus played an important role in shaping Canadians' understanding of food during war. Yet despite this ever-present discourse on improving individual food health, Canadians were also asked to ration important foods. Mosby notes the persistent fear, real or imagined, that shortages, despite the nation's vast food wealth, played in the public understanding of national food policy. At the helm of this discourse was the Canadian federal government, which entered the Canadian kitchen via price control and rationing. Mosby notes that such intervention was actually widely popular, not just for patriotic reasons. Canadians "came to see the wartime command economy as a possible alternative to an economic status quo that had brought only depression, unemployment, and poverty during the 1930s."[11] Although Mosby rarely references seafood, the fisheries industry worked within this larger context of state control and saw price control and rationing as potential aids to industry recovery by forcing more Canadians to eat more fish.

Mosby points to a "regime of economic controls" dictating food selection. The Wartime Price and Trade Board used advertisements, purchase books, peer pressure, and a coordinated system of voluntary rationing, even encouraging people to inform on those who did not comply, to control inflation and ensure that valuable food resources reached those in most need of them. All of this effort was based on the concept of "equity of sacrifice," in which everyone, whatever their income, contributed equally to the food needs of the nation and its allies. Despite occasional protests over rationing, this control of food and public commitment to high ideals of equity were generally successful throughout the war.[12] Canadians' commitment to reassessing individual and collective food habits provided the ideal context for the fisheries bureaucracy and industry to expand their pre-war efforts to move more Canadians to increase their seafood consumption. If successful in convincing Canadians to reduce their meat intake, then the government was right to assume that fish would replace the lost protein and fat. Thus eating fish could be presented as a patriotic step for Canadians to take in support of the war and also as an important part of the equitable distribution of food and nutritional resources.

Concerns over wartime food shortages, or at least the public belief that such shortages existed in North America as well as in Europe, and persistent calls to improve individual dietary habits added new plausibility to the Department of Fisheries' long insistence that the fishing industry needed to improve efficiency and increase the volume and reliability of its catch. In 1943, Donovan Finn, deputy minister of fisheries, wrote to George Akins of the Walsh Advertising Company, who had helped market seafood for the department and the Canadian Fisheries Association since 1936, saying that the department would not engage in any consumer-focused campaigns because demand for fish in Canada and in Allied nations now exceeded supply. The department's "resources and energies" were therefore taken up entirely by efforts to increase the catch.[13] Akins had worked for more than six years to increase consumption of fish in Canada with the objective of balancing the supply-and-demand economic equation that had long tilted to overproduction. Now it was production, not consumption, that must be increased.

Akins had regularly communicated with the Department of Fisheries throughout 1940–42, seeking funds for a continued advertising program. Now he retooled his approach. In a 1942 memorandum, he wrote that, after years of working on the problem of weak markets, he was positioned to provide "an analysis of the current situation in the Fishing Industry, with particular reference to the question of manpower, and the necessity for an increase in the catch." Proclaiming that food shortages in Britain demanded increased fish production in Canada, he asked the department to take actions to "stimulat[e] the fishermen now engaged in the Industry to an all-out effort." That stimulation would be done by uplifting their sense of pride in their livelihood and the role they played in the war effort. Akins proposed that the government fund a propaganda campaign targeting fishing communities on both coasts, boosting their morale to bring about increased effort and fish landings. A little over $125,000, he maintained, would be sufficient to inspire the nation's fishermen into "an all-out effort to increase catch."[14] Although the department did not in fact engage in this morale-boosting propaganda, the idea of increasing catch volume and reliability would come to dominate immediate postwar policy.

During the final years of the war, Allied governments designed food aid to control distribution of non-perishable food to aid Allied forces in their final push to defeat the Axis powers and aid civilian victims of war in Europe. The goal was to prevent possible postwar starvation and the political consequences that might emerge as a result of desperate need. Yet this altruistic goal was balanced by an equally strong desire to unload unmarketable food products. Rarely were products that could be easily sold in the United States or Canada shipped as food aid to Europe. Domestic consumption needs were calculated first and allotments made to ensure a profitable supply of the domestic market prior to any allocations of food for aid. In terms of food relief, hungry Europeans often got the least nutritious and least attractive products.

The immediate postwar global situation presented Canadian fisheries with opportunities and challenges. For more than two decades, the industry's primary focus was on increasing domestic consumption so as to free it from the uncontrollable variables of foreign trade. Now massive food shortages in Europe and the new Cold War desire to address potential European economic crises to ward off global socialism or communistic uprisings resituated Canadian fisheries policy decisions at least partially back towards international relations. This process began before the war's conclusion.

Nearly all of the scholarship on food aid, especially that related to international food aid, is rooted in political science. Within that field, the scholarship is largely divided into two periodizations, 1954 to 1972 and the post-1972 period, with the 1970s food crisis the main dividing point. Scholarship on the first period is largely defined by a realist approach to the analysis, emphasizing that economic and political self-interest defined food aid following 1954. The economic argument states that food aid was used as a means to dispose of surplus product in donor nations, thus aiding domestic food producers (mainly wheat producers), or as a method to develop potential foreign markets by manipulating food preferences in recipient countries. The economics of food aid also sought to facilitate economic development in recipient countries to inaugurate their entrance into a global food market. Alternatively, political realism suggests that food aid was largely motivated by political self-interest by using it as a tool of foreign policy to shape political development in the recipient country.[15]

Yet while realism and self-interest seem to have dominated food aid policy, especially between 1954 and 1972, they were failures in terms of policy execution. The food aid of those years was a poor method of disposing of domestic surplus and largely failed to create international markets for United States or Canada. Food aid in that period was also largely unsuccessful as a means to control political developments in recipient nations; in hindsight, it probably did more harm than good because it upset economic bases and often resulted in long-term food insecurity and social and political instability that proved especially problematic to the donor nations' political and economic objectives.[16] After 1972, there emerged what many political scientists view as a new humanitarian and multilateral approach to food aid. This new approach was dedicated to addressing international food security, defined as access to individual and community nutritional needs but also as equitable access to food resources, as well as the facilitation of food self-sufficiency in less developed nations. Peter Uvin argues in a 1992 article that since the food crisis of the 1970s, international food aid, especially that outside of the United States, revolved more around humanitarian justification than economic or political self-interest. Instead, Urvin suggests, "there exists a food aid regime which is predominately non-self-interest in nature" and "influences the international process to a significant extent." Urvin points to changing domestic law in the United States, Canada, and Great Britain, and to policies in international organizations like the United Nations and the European Economic Community that shifted the focus of international food aid away from political and economic self-interest towards securing international food security.[17] Historians, however, might argue that this was not a new humanitarian and multilateral approach but a renewed one. The post-1972 international food aid regime reflected ideologies similar to the pre-1954 philosophy, itself based on post-Progressive-era (1919–32) and New Deal (1932–39) idealization of social security, equity, and Franklin Roosevelt's declared "freedom from want."

The international food aid program that emerged immediately following the Second World War, one in which Canadian fisheries played a role, thus emerged from competing philosophical bases. The language of the debate that unfolded in Washington among international policy-makers associated with wartime food policies, such as those in the Combined Food Board, clearly rested much of their ideology on Roosevelt's promise of a better postwar world. Their language embraced the altruistic, humanitarian ideal of using the wealth of the United States, Canada, and Great Britain to feed the impoverished world and uplift poor nations in the same way that Roosevelt's New Deal had modernized much of the impoverished regions of the United States, mainly the American South. Lester Pearson, the key Canadian diplomat in Washington, was chair of the UN Interim Commission of Food and Agriculture, which oversaw international food aid between 1944 and 1954; he played a significant role in shaping the philosophical core of international food aid policy. Food aid was to be part of a larger modernization of efficient, globally planned food production, distribution, and consumption, while still embedded in the idealism of a global "freedom from want."

Some scholars have challenged this rhetoric. The new realities of an emerging Cold War and the political and economic necessities of fighting the spread of communism mitigated such idealism. In 1982, Harriett Friedmann used a Marxist theoretical structure to argue that US food aid policy in the immediate postwar period was specifically designed to facilitate international capitalism by shifting more the world's population away from food production towards market trade: "The extension of commodity relations to the food supply became an intrinsic part of the project of capitalist industrialization and shaped its course." Western food aid was thus, according to Friedmann, used to facilitate industrialization by reducing agrarianism and so generating more consumers of food and other products in an international capitalist marketplace. In this view, US food aid was designed not to feed the world's hungry masses but to restrict trade with socialist states, undermine domestic food production (self-sufficiency) in less developed nations, and build international capitalism.[18]

Yet Friedmann also recognized entrenched economic realism in 1940s US food aid policy. Delving far more deeply into the historical context of food aid than most scholars, she showed how the rhetoric of food aid drew from Roosevelt's domestic New Deal food policy but did so from an economic realist rather than a social idealist perspective. The powerful farm lobby that had emerged within US politics during the Great Depression retained important lobbying power over postwar policy and sought to use food aid as a means to dump a food surplus that had increased during the war years. Friedmann identified the roots of both economic and political realism, which the broader scholarship typically reserves for the post-1954 period, in the final years of the Second World War and even in the pre-war political philosophy of modernization that sought to balance production and consumption in an industrialized state.[19] J.R. Tarrant came to similar conclusions in 1980: "Although there are also more altruistic and humanitarian reasons for food aid, particularly during major famines and natural disasters, surplus disposal probably remains the central objective of food aid or, at the very least, no food aid would exist without a surplus." Like Friedmann, Tarrant took a deep historical approach to the study by identifying direct links between domestic food policy during the Depression and postwar international food aid policy.[20]

More recently, Carmel Finley has addressed the impact of postwar foreign policy on US fisheries policy and the influences that policy had on shaping Canadian, Japanese, Russian, and other states' marine resource management. Finley shows that postwar American fisheries policies "was built on a foundation that saw man as playing a positive role in harvest and believed that populations had surplus production that could be harvested. Fish could be sustained at high levels into perpetuity."[21] Her book focuses on the Maximum Sustainable Yield concept and argues that the construct was not scientific but political. US fisheries policy sought to address the politics of postwar food policy more so than the idealized version of Franklin Roosevelt's "freedom from want."[22] Finley addresses conflicting views of US postwar food policy as a product of New Deal–era idealism or Cold War–era realism: "The internationalists supported the idea that all countries should have free access to the world's resources. The 'realpolitik' faction, led by a cadre of legal scholars and backed by the fishing industry, argued that the United States had to act in its own best interests to preserve the fish, regardless of the international implications."[23]

In many ways, Canadian fisheries policy after the Second World War fits this same paradigm, but it is important also to remember that fish was part of the new international food order. Fisheries industry leaders in Canada sought to protect their domestic resources from foreign interference, while people like Lester Pearson, chairman of the Interim Committee for the UN's Food and Agriculture Organization (FAO), and Donovan Finn, Canadian deputy minister of fisheries and chairman of the fisheries committee of the FAO, wanted an efficient and orderly global food regime that required sharing food-production knowledge with developing nations.[24] In terms of the fisheries, this regime called for advances in fisheries productivity in the southern hemisphere while seeking fisheries conservation in the northern hemisphere.

Finn held the fisheries committee chairmanship within the UN's Food and Agriculture Organization and was one of the few fisheries scientists or policy-makers, along with Japan's Tomonari Matsushita and Britain's Michael Graham, Sidney Holt, and G.K. Kesteven, to resist American insistence on basing global fisheries policy on the theory of Maximum Sustainable Yield.[25] Nevertheless, Canada abstained from the all-important vote at the International Technical Conference on the Conservation of the Living Resources of the Sea held in Rome in 1955. This meeting, as Finley argues, established Maximum Sustainable Yield as the guiding principle for international fisheries management. It allowed established fishing nations to broaden their geographic reach and exploit the marine resources of less-developed nations. Canada potentially fell into both camps; it wanted to protect Atlantic and Pacific resources from European, Russian, and Japanese fleets but was also interested in taking its historical fisheries knowledge and engagement globally to the internationalization of marine resource exploitation. Where exactly Canadian policy fell in this dichotomy is hard to pinpoint, but for the purposes of the present study, it is clear that Canada would lose international customers if other nations built or rebuilt fishing fleets.[26]

Canadian fisheries involvement in postwar global food aid evolved from two sources. The first was via wartime food control conducted by the Combined Food Board, an inter- and intra-governmental organization made up of American, British, and Canadian representatives. This board focused on organizing the shipping of food products identified as being in short supply. The program was a grand undertaking designed to create global, or at least Allied, control of food distribution. The board determined total global production of thousands of food products, assessed total global need for them, and assigned allotments intended to provide equitable access to needed food commodities independent of a nation's or its people's ability to pay. It reflected much of the early New Deal's optimistic views of planned economics.[27]

The second venue through which Canadian fisheries participated in postwar food aid was a series of relief programs including the United Nations Relief and Rehabilitation Administration (UNRRA), the United States–based European Recovery Program (Marshall Plan), and the United Nations Food and Agricultural Organization (FAO). These agencies also sought more global control of food supply and were thus similar to war-related food programs like the Combined Food Board, but they focused more on "relief feeding" of civilian populations affected by the war. It is important to note that relief feeding, the main focus of the UNRRA, was intended to be temporary, whereas the main goal of postwar reconstruction under the Marshall Plan, for example, was to restore Europe's capacity to feed itself. Yet Canada maintained that Europe had not been self-sufficient in food production in pre-war years but had instead relied heavily on North American food production. Making Europe self-sufficient in food production would mean Canada would lose important pre-war customers. This would be particularly problematic for the fisheries, which stood to lose key Mediterranean and British imperial markets that were especially important to the nation's dried fish industry.[28]

In both wartime food-supply control and postwar relief feeding, food aid could provide Canadian fisheries and the nation's food industry more generally with an outlet for overstock food products. The programs rarely addressed recipients' nutritional needs. A consumer-focus history of fisheries food aid exposes a consistent problem of Canadian fisheries throughout the twentieth century: overproduction, based on poorly calculated consumer demand. The US goal to restore European fishery economies meant a collapse of European markets for Canadian fish, and, once again, the fisheries industry and the Department of Fisheries had severely overestimated consumer demand when they apparently assumed that wartime food needs would continue well after the war. The drastic decline in 1947 in government purchases of seafood for European aid led to yet another glut in production.

On 9 June 1942, Franklin D. Roosevelt and Winston Churchill had created the Combined Food Board and placed US secretary of agriculture Claude Wickard and the head of the British Food Mission, Thomas Henry Brand, at its helm. The objective of the board was to organize the utilization of all food resources, including food production material and machinery, for all UN members. The specifics of this ambitious objective would be worked out by subcommittees on specific food resources, which themselves would be staffed by representatives from member nations. The goal was to internationalize food resources to prevent costly food competition between Allied nations, which could result in inflation and misallocation of valuable food resources. One of these subcommittees was the Fish and Fish Products Committee, on which Finn served as the Canadian representative. Ray Gishue, of Newfoundland, served as the committee's chair.[29]

Upon hearing of the creation of the Combined Food Board, Canada quickly created an interdepartmental Food Requirements Committee to oversee Canada's contribution to Allied food efforts.[30] In a September 1942 communication, Prime Minister Mackenzie King noted the central role Canada had to play in supplying Allied food needs: "The complete mobilization of resources for the Canadian war effort makes it necessary to provide for continuous consideration of Canada's undertaking to export foods to the United Kingdom and other United Nations."[31] Yet Great Britain did not feel that Canada needed an equal voice as an executive member of the Combined Food Board; it expected Canadian representatives to work primarily at the subcommittee level. In a September 16 meeting with Mackenzie King, Brand noted that the Combined Food Board was responsible only for overseeing products in short supply and for coordinating shipping of supplies to Europe. Neither task, he concluded, required Canadian leadership.[32]

Mackenzie King objected. As a world food provider, and one particularly important to Allied objectives in the war, Canada should play a leading role in North American supply of food to Europe: "Canadians took a great interest and pride in their achievements in the supply of food."[33] Mackenzie King's focus on Canadians' patriotic calling was quickly balanced out by the more economic considerations in Canada's food industry as voiced by the minister of agriculture, James Gardiner: "Canadian producers were nervous lest, under the Lease-Lend agreement with the United States, they would be deprived of the opportunity to supply the United Kingdom, and [at] least a Food Board composed only of representatives of the United Kingdom and the United States would tend to arrange for the procurement in the United States under the Lease-Lend procedure of commodities which Canada was ready and anxious to supply."[34] Malcom MacDonald, high commissioner for the United Kingdom, quelled Gardiner's fear, noting that an immediate communication from Great Britain could confirm an "assurance previously given that the United Kingdom would draw needed supplies from Canada in preference to the United States when supplies from both countries were available." As long as Canadian food producers retained essential customers, the Canadian government seemed content with the arrangement, for the time being.

During the first months of its existence, the Combined Food Board faced difficulties in Washington, largely because of internal fighting between the US Department of Agriculture and the US War Food Administration over who controlled wartime food planning in the United States.[35] Within this context, British representatives changed their minds concerning Canadian representation on the executive committee and pressed the United States to allow for this third, potentially calming voice. According to Canadian representatives in Washington, the British believed "a Canadian representative could help greatly to improve the operation of the Board."[36] In a 28 October 1943 press release from the Combined Food Board, both Churchill and Roosevelt issued a public invitation to Canada to join: "Canada's contribution to the war effort in the whole field of production and the strength which she has thus lent to the cause of the United Nations is a source of admiration to us all. The importance of Canadian food supplies and the close interconnection of all North American food problems makes it appropriate and desirable that she should be directly represented as a member of the Combined Food Board sitting in Washington."[37] Mackenzie King named Gardiner as Canada's representative, and Canada took on an equal role – at least on paper – with the United States and Great Britain in overseeing an ambitious plan to control the totality of Allied food production, distribution, and even consumption.

The extent and depth of the Combined Food Board's analysis and direction of the global food supply should not be underestimated. Its numerous subcommittees assessed total global supply, determined regional and national needs, and allocated amounts for distribution, all of which was enforced by cooperating national governments, which could use the wartime crisis to control food production, distribution, and consumption within their nation. What the Combined Food Board sought was a global food system of planned food production that embraced control over not just food commodities but also supportive food products such as fertilizer and machinery. For example, in a press release of 30 July 1943, the board announced a system that oversaw 90 per cent of Allied production of fats, oil, and oil seeds and allocated specific distribution of those resources to ten consuming areas: the United Kingdom, the United States, Canada, Australia, New Zealand, the Soviet Union, the Middle East, South Africa, and neutral European nations.[38] This control was not only broad at a global/macro level but also in depth at a local/micro level. In a meeting on 9 August 1943, for example, the Subcommittee on Dairy Products in their calculations for global distribution estimated the amount of butter needed for cooking in Trinidad.[39]

Fish, and especially canned seafood, was an important focus for Canadian representatives on the board. Although a minor commodity in global food supplies, and even a minor commodity within Canada's contribution to that global supply, the fish products allocated by the board involved the vast majority of Canada's fisheries resources, covering some commodities in their totality. The Canadian interdepartmental Food Requirements Committee communicated with the Combined Food Board in Washington and related all fish and seafood requests to the Canadian Department of Fisheries, which arranged all purchases of Canadian fish products for the Combined Food Board, and for future relief programs such as the UNRRA or the US Marshall Plan. Government oversight of the nation's fisheries was now all-encompassing.

The impact of Combined Food Board purchases on Canadian fisheries should not be understated. G.H.S. Pinsent, of the United Kingdom Food Mission, noted in January 1943 Britain's growing need for Canadian fish. Canadian fresh and frozen fish were in demand for Britain's fish-and-chips canteens, which "play an essential part in wartime supply of the U.K. working class," and the need for canned seafood continued. Pinsent wanted 6,000 long tons of fillet and requested that Canada redirect some of its exports from the US market to Britain. Finn, however, disagreed, pointing out the economic realities of the industry: "The United States was the main permanent market for this type of Canadian fish and ... high United States prices were the chief incentive for present production." Without profit incentives, the industry would not produce enough fish to provide for global wartime demand. Given the long history of overproduction in this particular sector, and its continual reliance on government purchase, one might question Finn's conclusion. Nonetheless, he allowed only the allocation of 4,000 long tons of fillet for the British market.[40]

Other fishery sectors without ready profitable markets in the United States and Canada, however, became increasingly dependent on government purchases for foreign food aid. The canned salmon industry, which controlled a major share in Canada's total seafood production, seemed nearly totally dependent on wartime government purchases. In 1943, Finn expected British Columbia to produce 1.5 million cases of canned salmon, of which 1.2 million cases would go to the United Kingdom (less than the British had requested), and another 80,000 to the Canadian Red Cross, ships' stores, and the Munition and Supply Department. This left only 220,000 cases for domestic consumption.[41] In another example, in 1948 the British Ministry of Food purchased 100,000 cases of canned herring at a rate 25 per cent higher than going market value. The total pack for Canada that year was only 250,000 cases; the canned herring sector was able to sell 40 per cent of its entire year's production at profitable rates via a tightly controlled government marketplace.[42]

British purchases of canned fish added significant capital to the industry. In 1943 alone, Britain spent $16,550,000 on canned salmon and $12,800,000 on canned herring.[43] Expenditures for the Canada Mutual Aid Board, which provided supplies to Allies on a half-credit, half-cash basis, also reflected the tendency to focus on canned fish.[44] For the 1943–44 fiscal year, the Mutual Aid Board spent $9,239,000 on canned salmon, $10,140,000 on canned herring, and only $1,343,000 on frozen fish. Thus, the positive impact of war aid was felt particularly strongly in the preserved fish industry.[45]

The Canadian Red Cross was another major purchaser of Canadian fish products, and it too focused on canned fish. Percival H. Gordon, chairman of the executive committee for Canadian Red Cross, wrote to Finn in July 1941 that his organization planned to pack 40,000 prisoner-of-war parcels per week and requested a half-pound tin of salmon for each package.[46] Although not nearly as large as the volume going to Great Britain, this was still a significant amount for the industry. In 1942 alone, the Canadian Red Cross took 20,000 cases of canned salmon and 20,000 cases of canned pilchard from British Columbia and 45,000 cases of sardines from New Brunswick.[47] As the war progressed, the need for prisoner-of-war parcels obviously increased. In 1943, the Canadian Red Cross packed 100,000 parcels per week and requested from the Department of Fisheries an allotment of 52,000 cases of canned salmon and 53,000 cases of sardines.[48] In 1944, the federal government or international aid organizations such as the Red Cross purchased the entire production of Connors Brothers of Black Harbour, a company long involved in marketing their product to domestic consumers.[49] That year not a single sardine from the nation's only sardine producer on the Atlantic went into the private domestic marketplace.

At the helm of this increasingly managed seafood industry stood the Department of Fisheries. It determined potential domestic demand for any given product and then assessed total foreign aid demand, via the Combined Food Board and organizations like the UNRRA or the Canadian Red Cross, as well as total Canadian Forces demand. Only after these calculations did it issue export permits to allow companies to sell products to foreign buyers not associated with wartime or postwar food needs. Such tight scrutiny of the free market, it insisted, was a "measure of control to secure the maximum production of certain fish commodities which the Department is under obligation by agreement to supply to the United Kingdom Minister of Food and to protect the requirements of the Canadian consumer."[50] Direct control over fisheries products most affected less-perishable products, mainly canned seafood. Virtually no export permits were issued during the war years for Canada's canned seafood other than canned lobster, a luxury item not often associated with relief feeding. More generally, however, by the war's end, Canadian seafood consumption was roughly divided between domestic markets, export to the United States, and exports to all other nations, with the United Kingdom consuming about two-thirds of this last category. A larger percentage of total export, however, was in the form of food aid to the United States, for redistribution through the Combined Food Board or to Great Britain. With most canned fish, and nearly all canned salmon, going to the United Kingdom, the supply to the domestic market was mainly fresh and frozen fish.[51]

Beyond just wartime needs, the larger geopolitics of postwar food aid also affected Canada's seafood industry. On 9 November 1943, the United Nations had created the UNRRA "to cope with an emergency, to make available the food and supplies considered essential to prevent starvation and disease." Nations were to contribute 1 per cent of their national income to the program, 10 per cent of it to come in the form of cash and the remainder in credit for purchases in that country of needed relief and reconstruction supplies. Canada's contribution was expected to be somewhere between $80 million and $90 million, while the US contribution would be in the range of $1.3–$1.5 billion.[52]

In 1944, Canada contributed just over fourteen million pounds of fish to the UNRRA, above that taken by the Combined Food Board, the Canada Mutual Aid Board, and the Red Cross. While places like the British Food Mission and the Red Cross received canned salmon, the UNRRA got much lower-grade products like canned codfish, haddock, hake, and mackerel.[53] Other products like ground pilchard loaf also made it into the UNRRA purchases, but the agency did reject products that it believed would be unacceptable to potential relief recipients, such as sardine spread.[54] The UNRRA insisted that relief food had to have adequate nutritional quality, assured keeping qualities, acceptability by recipient countries, and a satisfactory price.[55] It was the last requirement, however, that mattered most, and, as such, products like ground pilchard loaf found their way from Canada to Greece.

The American European Recovery Plan, commonly known as the Marshall Plan, also sought a balance between protein and price that tended to favour lower-grade products. Silver hake, river herring, sea herring, and lower grades of canned seafood had the most potential for purchase under the Marshall Plan.[56] Some products were, however, rejected. H. Burhoe, manager of J.W. Windsor Co. Ltd in Charlottetown, for example, was upset that the government rejected his supply of chicken haddies, "which had only been degraded because of the presence of a few worms."[57] Senator William Duff of Lunenburg was likewise disappointed that the government rejected fish cakes from his province. Stewart Bates, acting chairman of the Fisheries Price Support Board in Ottawa, explained that the fish cakes were 80 per cent potatoes and thus did not contain much protein, but Duff felt they were good enough for "starving Europeans."[58]

Food aid constituted only one minor component of the UNRRA. Its main focus was on rebuilding Europe's productive capacity by providing nations devastated by the war with machinery, technology, and production-based commodities like fertilizers and seeds. This emphasis proved problematic for the Canadian fisheries, which had become accustomed to the growing demand for its products. If Europe were to become self-sufficient in fisheries production, Canada would lose valuable customers. Nonetheless, the UNRRA's Expert Panel of Fisheries, a subcommittee of the Committee on Agriculture, insisted that rebuilding Europe's fishing fleet was the primary goal of fisheries relief and rehabilitation. In an October 1944 report, the panel concluded, "In view of the fact that the expansion and resumption of fishing results in supplies of food of high nutritive value becoming available immediately, the Expert Panel of Fisheries urge that deliveries of Gear and other requirements should be given the same rank of urgency as deliveries of requirements for rehabilitation of agriculture in Enemy Occupied Countries."[59] The UNRRA focused primarily on restoring Europe's capacity to feed itself and creating regional self-sufficiency.[60] In 1944, the UNRRA estimated it would require about 109,000 metric tons of fish of all kinds from all sources. In comparison, it estimated it would need 207,840 metric tons of maize and 216,150 metric tons of sugar; wheat, at 1,531,325 metric tons, made up the single largest predicted purchase.[61] Canadian fisheries, therefore, had little to add and little to gain from the UNRRA.

The Canadian Department of Fisheries continued to control allocations to the domestic market, Combined Food Board, Red Cross, British Food Mission, Canadian Mutual Aid Board, and UNRRA throughout 1946, but by then government purchases of fish began to decline; by 1947 they had all but disappeared.[62] The sudden loss of a major purchaser was a great blow. Canada's fishing industry entered the 1946 season with the full expectation of high demand from government purchasing programs and had thus invested capital and labour in extracting large volumes of fish from the seas. The UNRRA's cut in its intended purchases for the 1945–46 fiscal year by $6.2 million resulted in yet another glut in the global fisheries market. Canada's canned fish sector alone faced a three-million-pound overstock. The Price Support Board purchased the overstock from the industry, but the UNRRA shortly announced its desire to cancel its full order of thirty million pounds of canned fish for the 1946–47 fiscal year.[63]

The Price Support Board had been set up by the 1944 Fisheries Price Support Act to purchase any unsold fish food product produced by Canadians for war-related purposes. The idea was to give producers some reassurance they would find markets for their product and thus encourage maximum productivity during a time of assumed food shortage. After the UNRRA sought to be released from its contract, the board ended up purchasing the entire pack of 1.6 million cases of Pacific herring in the 1946–47 fiscal year.[64] Its ability or willingness to continue these purchases, however, was limited, and the prices paid were significantly reduced.[65]

In the summer of 1945, some UNRRA officers were still claiming global food shortages even as others argued that the organization's poor food management and inequitable distribution, not total supply, were the main problems.[66] In Great Britain, the United States, and Canada, studies found that wartime food supplies had actually improved and, in many cases, resulted in healthier diets. When the head of the UNRRA, Andy Cairns, sought to encourage shipping frozen fish to Italy, even though Italy lacked the storage capacity for it, Canadian representatives in Washington suggested the agency was somewhat mismanaged and overzealous it its global quest for food. Cairns, they advised, "is a person around whom much controversy reigns, but one whose most undeniable characteristics are enthusiasm and optimism. Perhaps, rather than a realistic appraisal of what these countries could properly receive in the way of frozen fish, Cairns once again allowed his enthusiasm to get food to Europeans to guide his decision."[67]

The intention of the American-based European Recovery Program (ERP) was also to rebuild Europe rather than simply feed it. Its focus was on reconstructing the Continent's fishing capacity rather than the shipping of North American fish over in the form of food aid. Within this context, the chief economist for the Department of Fisheries, Ian McArthur, felt "most pessimistic regarding the prospect for Canadian fishery products being included in the Program."[68] The Newfoundland colonial government had equally grave concerns around the Marshall Plan's intended goal to rebuild Europe's fisheries. The governor of Newfoundland, Gordon Macdonald, predicted in a letter to Philip J. Noel-Bale that the Marshall Plan would be "a death blow to Newfoundland's long established trade in salt codfish with Europe."[69] The program, as Macdonald understood it, proposed bringing to an end a two-hundred-year Newfoundland history of providing food to Europe. By 1947, most of the Allied leadership had moved away from the idea of relief feeding, and, drawing from what they saw as lessons from the post-First World War period, instead concentrated on building a global food management program that would ensure equitable access to food resources and a nutritional base to ward off a potential global hunger or malnourishment crisis. This emphasis led to the abandonment of UNRRA and ERP and the creation of the new, more robust, and more permanent FAO.

Franklin Roosevelt's declaration that all peoples deserved "freedom from want" drove much of the United Nations' early assessments of a global food management program. In October 1942, a group of international social activists, seemingly as much driven by Roosevelt's Depression-era rhetoric about social equity as by wartime realities, issued a draft memorandum outlining a potential UN declaration of a "freedom from want of food." This document, which Lester B. Pearson quickly relayed to Ottawa, would become the philosophical basis for the FAO.[70] It was Pearson who would chair the interim committee in the formation of that organization. Given that people like Dr Hazel Stiebeling and Dr John Orr, both pre-war social activists who associated malnutrition and hunger with poverty and demanded government welfare programs to eradicate inequality, played leading roles in the interim committee, it is unsurprising that the first documents from this organization were laced with strong social justice rhetoric.[71]

The initial draft proclaimed, "Freedom from want of food must be given high priority in the actions taken to fulfill the pledges of the United Nations." The draft also articulated the necessity of governments creating programs to facilitate equitable access to nutrition, and it celebrated wartime rationing systems for ensuring regular supplies of food while preventing inflation and, more importantly, for creating a context for a more equitable access to food of high nutritional value. The experience laid the foundations for a potential postwar global food management program that would ensure increased food production efficiency while allowing for globally equitable distribution. The manifesto concluded, "The cumulative effect of the mass of evidence available from many countries is to establish the truth that if food, in sufficient quantity and variety were made available to all income groups this would eliminate the worst effects of poverty and do more than anything else to raise the standard of living."[72] This globally planned food regime as expressed in early drafts revolved around key social justice concerns rooted in pre-war assessments of inequality in the United States, the United Kingdom, and Canada. The draft became the basis of the June 1943 conference on food equity at Hot Springs, Virginia.

In January 1944, the Interim Commission on Food and Agriculture began the exhaustive work of building a postwar globally managed food regime. The defining focus was to study existing food resources, determine how to increase food production via improvements in efficiency, and find methods for more equitable distribution of those resources. The commission included several steering committees, including ones on agriculture, nutrition and food management, forestry, and fisheries. Donovan Finn, Canada's deputy minister of fisheries, was named as the Fisheries Committee chair and had a seat on the review panel that would organize the commission's final report. His agenda for the FAO Fisheries Division program was to "examine the possibilities for expansion of present fisheries, since in many parts of the world known marine resources could without harm be exploited much more intensively to supply food for human beings, feed for livestock, and material for industry. It should encourage systematic exploration for virgin marine resources and evaluate the possibilities for their development. It should encourage the setting up of additional research laboratories to study biological, economic, and technical problems related to the fishery industry throughout the world."[73]

Despite a history of overproduction, the FAO set the fisheries committee the task of increasing production. Finn, however, was concerned that this direction was rooted in poor data on consumption. In a special meeting of the Interim Committee's Review Panel in August of 1944, he replied to questions concerning statistical data of global demand for fish products: "There is an almost total lack of information on pre-war consumption levels. The Combined Food Board's Report on consumption levels in wartime was possible because of wartime controls. This raises a real problem affecting not only fish consumption, but all food consumption, [as] when controls disappear, consumption statistics will again become very inadequate, unless some way is found of carrying over into peace time some of the devices for records and information."[74] During his participation in these meetings, Finn mainly stressed the lack of data and the inability to complete reports as detailed as those of agriculture because of limited investigation into fisheries supply, productivity, or consumption. One of his main agenda items was to create a global fisheries database to capture nations' fishing efforts and thus be able to construct policy based on more accurate information.

In the end, whereas the committees on agriculture, nutrition and food management, statistics, and forestry all produced many thousands of pages of documentation, the Committee on Fisheries came to few conclusions. In a November 1944 meeting of the Review Panel, the chair noted the absence of a draft on fisheries to review as it constructed an outline for its final report. The Nutrition and Food Management Committee had submitted a 124-page draft. The Agriculture Committee, probably the most robust of the group, had submitted numerous subcommittee reports on such topics as "Subsistence Agriculture in Newfoundland" and "The Grasslands of Southern China." Yet despite the Fisheries Committee's failure to produce a supporting report, the Review Panel concluded, "A brief review of the present situation shows that certain waters have been over-fished, others have been largely neglected and everywhere there is gross wastage of the landed product both in marketing and processing."[75] This bold yet vague sentence suggests problems of overfishing, under-fishing, and waste of product due to poor handling or low consumption. Nonetheless, the committee championed the idea of increased scientific exploration of the fisheries and a better communication of that knowledge to undeveloped regions of the world.

In January 1945, still apparently without a finished report from the Fisheries Committee, the Review Panel produced a first draft of its summary for the Interim Commission on Food and Agriculture. The panel identified the central problem as a need "to increase the total output and to make better use of the existing supply." Yet it also argued that the fisheries had historically been "treated as a mine rather than as a cropping ground and the fish have been carelessly taken without regard to continuity of production."[76] Its rhetoric of increased productivity was thus partially balanced by a call for international conservation efforts and a need to understand variables of supply-and-demand across time and space. The report noted that, in general, the northern hemisphere had been overexploited and the southern hemisphere underutilized. The result was a global imbalance, or inequity, of fish use. The panel also expressed concern that "the immediate danger after this war, as after the last, will be an orgy of over-fishing because of the shortages of protein foods."[77] This language echoed other sections of the Interim Commission's report, which also focused on equitable access to the world's food resources and a need to create global balance of supply-and-demand to prevent inefficient and dangerous fluctuating periods of overproduction and scarcity. The final report from the Agricultural Committee, for example, warned, "there are in our view serious dangers of maladjustments between output and consumption so that positive steps are needed to maintain the two in balance."[78] Clearly, the pre-war experience of glutted food and agricultural markets had as much influence on the FAO's initial foundational philosophy as the then contemporary fear of wartime food shortages and postwar famine with all of its potentially ruinous political and social effects.

As the UNRRA finished its operations in 1947–48, Canadian fisheries once again faced limited demand. Ian McArthur, chief economist for the Department of Fisheries, recognized this and sent the deputy minister an assessment of the impact of relief feeding on the nation's fisheries: "I feel that the post-UNRRA program did little more than temporarily relieve the situation. Little of a constructive nature was accomplished other than some encouragement to quality production." McArthur summarized how dependent the industry had become upon government purchases: "Thus, if the Government is not buying at any particular time, many canners close up, the fisherman loses his market, and the income of the industry is reduced. If some agency were able to provide the necessary financial strength to permit the building up and holding of temporary surpluses, production could continue uninterrupted."[79] As in the post–First World War period, industry and government scrambled to find an outlet for surplus supplies and recognized that little had been done during the boom years to establish long-term economic sustainability.

The years following the Second World War began much like those that followed the previous war. In 1948, the industry corporate cooperative, now relabelled the Fisheries Council of Canada, once again pushed the Department of Fisheries to mobilize aggressive advertising to increase domestic consumption of fish as a means to relieve the industry's economic stagnation and the depressed livelihood of fishermen and shore workers.[80] The focus of this would-be campaign, however, was significantly different from the pre-war one. Clive Planta, secretary-manager of the Fisheries Council of Canada, stressed the importance of improving the quality of the product before pushing to increase consumption: "It is generally recognized in the industry that until there is more evidence of a continuity of quality products being available to consumers with protection throughout the channels of distribution from the time of catching until sale at retail level, it is impracticable to expect the Government to spend public funds for advertising to encourage increased domestic consumption." Planta's emphasis was a substantial shift from pre-war assumptions about the potential impact of marketing. Instead, what he and other indus-

try leaders proposed was more of a public-relations campaign
designed to educate consumers about the nation's fisheries, including
everything from geography, biology, and history to "dramatic inci-
dents" from fisheries folklore. This campaign would lay the seeds for
future consumption once the industry had improved the product.[81]
The Fisheries Council of Canada approached the Department of Fisheries again in 1949 with the same idea. Neither request, however, met
with success.

Quality control became the Department of Fisheries' primary focus
following the war. Its own Committee of Advertising, chaired by
H.F.S. Paisley, who had overseen the department's 1936–40 campaigns, concluded in 1947 that any advertising would be "ill-timed at
this junction." Efforts to increase domestic consumption of fish would
fail if quality-control measures were not first enforced to ensure the
product reached the customer in a far superior state than what was
being produced.[82] Two years later, O.C. Young of the Fisheries
Research Board of Canada also maintained that "the unfavourable
consumer reaction resulting from reduced quality cannot be estimated,
but the low per capita consumption of fish in Canada can be placed
at the door of poor quality."[83] In his inaugural address, the new minister of fisheries, Robert Wellington Mayhew,[84] again emphasized the
need to get "the highest quality, inspected foods to the housewives."
Reaching that goal would necessitate improved and expanded inspection of fish and processing plants, consolidation of fishing operations
to improve efficiency, faster and more reliable transportation, better
storage in wholesale distribution centres, and improved presentation
in retail operations.[85]

By the late 1940s, more government and industry leaders recognized the need to expand the government's inspection authority to
improve the quality of the product across the industry. In a 1952
report to the prime minister, the Department of Fisheries continued
to insist that industrial modernization must precede improved market conditions: "Quality depends on how fish is handled in boats, in
plants, in the selling trades, yes, and in the kitchens of restaurants and
homes ... From sea to table is a long assembly-line: each part must do
its own job well, but each has to fit into the other. Any programme
must cover each of these steps, and try to provide for the industry the
conditions that will allow it to do each of the jobs better."[86] Much of
this focus on quality control might suggest that the fishermen's union
had finally had their voices heard by the Department of Fisheries, yet

the department rejected the fishermen's real concern: the banning of trawlers. Instead, it argued that industry modernization was central to product quality and would come with increased concentration of economic power and increased utilization of trawlers and other advanced fishing and processing technologies.

Beginning in 1951 and continuing until 1965, the Department of Fisheries ran its own public-relations campaign independent from the industry. It was strikingly different from that of 1936–40; in the first place, it emphasized quality. In a 1955 summary of the program, the deputy minister, George R. Clark, explained that "the limited advertising budget of the Department is used to stress the need for top quality as a prime requisite in increasing the consumption of fishery products." The main targets were younger audiences "who have not yet become set in their eating habits."[87] Also, much of the advertising was directed at the industry rather than the consumer. A 1962 advertisement showed a woman's gloved hand reaching for a package of frozen fish and declared to the industry: "Every time a Canadian Housewife reaches for fish – she puts money into *your* pocket!" The message urged the industry to ensure that the housewife would be satisfied with the product's quality and thus be motivated to make additional purchases.[88] This campaign seemed to shift responsibility for industrial recovery away from Canadian consumers who needed to buy more fish and towards the Canadian fishing industry who needed to produce better fish – something Evelene Spencer had told the industry in her 1934 article. Thus, much of the department advertising budget was spent in fishing communities rather than major metropolitan centres as in the 1936–40 campaigns.[89]

Secondly, and perhaps more importantly, this public-relations campaign went beyond advertising fish or seafood to underline the work of the department itself. A 1961 newspaper campaign text, for example, claimed the department "engages in the conservation and expansion of fish populations." The advertising copy seemed most interested in explaining to Canadians the work of the Department of Fisheries and how that work was beneficial to the industry, workers, and consumers. The department, the advertisement continued, "also inspects and promotes fishery products, and stimulates those engaged in the industry to keep abreast of technological developments."[90] A 1965 advertisement stated, "Assuring an adequate supply of fish is a big job the Department of Fisheries does in a hundred ways," and "the quality of the fish you buy and serve is a consequence of high standards and rigid inspections of plants and equipment, another Department of Fisheries responsibility."[91] These ads presented the department as a kind of consumer protection agency dedicated to ensuring Canadian customers got the highest possible grade of fish. By improving its public image, especially as it related to assuring quality and safe food for consumers, the department hoped to stimulate increased consumption by addressing the reputation of fish as a low-quality food product. In many ways, the advertising justified the department's new role as an inspector of fish products. It also created a separate Home Economics section, which staffed demonstration kitchens in Halifax, Montreal, Ottawa, Toronto, Winnipeg, Edmonton, and Vancouver with home economists who gave regular cooking demonstrations and appeared on local radio and television shows to popularize eating fish and teach consumers how to choose and best prepare it for family meals.[92]

Thus, following the Second World War, the Department of Fisheries insisted that industry recovery, stability, and growth must be based on quality of production. It pushed for and eventually got new regulatory authority for an expanded inspection system that included all levels of production, not just the landing of the fish but also its processing, distribution, and wholesaling and retailing. In an government-industry meeting in 1947, the new deputy minister, Stewart Bates,[93] told industry leaders, "All plants processing frozen packaged fish will have an inspector in them for the purpose of assuring that only quality fish is used and that at least the quality of the fish out for packing is equal to the quality that is presently selected for the fresh fish trade." Increased consumer demand was possible, he believed, but could only come after assuring high "quality control," which was a "long-range undertaking."[94]

This shift away from quantity of consumption towards quality of production did not necessarily result in a re-evaluation of overall production and calls to manage that production for long-term sustainability. It is important not to mistake the department's post-1947 calls to increase quality as an endorsement of pre-war calls by fishermen's unions that had also focused on improving quality by banning trawlers and the trash fish they produced. In 1947, for example, B. McInerney, head of the Salt Fish Administration within the Department of Fisheries, estimated that total salt fish production that year would top forty million pounds, "a high figure." Although both the salt fish and fresh and frozen markets were already glutted, McInerney explained this high level of production: "Capital investments in the implements for catching fish and shore processing plants cannot be readily liquidated for diversion to other pursuits and therefore must be kept working to the greatest extent possible for its own protection."[95] Despite low market demand, trawlers would have to increase their fishing capacity to offset decreased retail per unit value. Trawlers kept fishing even though customers were not buying. No one seemed concerned about the increased pressure on the environment. This laid the ground for future collapse. The fisheries collapse of the second half of the twentieth century was, at least in part, the result of the fact that, during the first half, no one seemed to have any idea how to balance production levels with consumer demand.

Selling the Ocean:
How Marketing Goals
Affect Resource Sustainability

A consumer history of seafood in Canada can add insight into probing questions on why the fisheries failed. Whereas most contemporary observers tended to point to variables in production, uncontrollable vagaries of nature, the limited capacity of existing technology, or the lack of good scientific knowledge on supply levels, this work points to the near total failure within the fisheries industry and the fisheries government bureaucracy to correctly predict or control consumer demand. Although production fluctuated at sometimes extreme rates during the twentieth century, the market was more often over-supplied than under-supplied, with the exceptions of wartime demand. The persistently glutted market led to perpetual loss of retail and wholesale value in relation to cost of production or unit of effort. The declining returns on investment forced large operators to increase overall production to offset decreased per-unit value. Trawlers kept on fishing even though they could not dispose of their catch at appropriate rates to compensate either labour or management, only accentuating the problem by further glutting the market and driving down value.

Few foresaw the long-term ramifications of increased fishing effort on the environment that sustained that resource economy; most still blindly believed in the fisheries' inexhaustibility. The result was an oversupplied market that could not profitably compensate labour and invested capital, creating short-term regional economic recession as well as a collapsing resource that set the stage for long-term disaster. Although the retail value of high-valued fish commodities such as fresh frozen fish fillets, canned sockeye salmon, and lobster remained high, the overall wholesale value of seafood products plummeted as raw product supply increased and producers struggled to find domestic or foreign markets for it. Throughout the first half of the twentieth century, fish producers turned to government assistance to break the cycle of imbalance between production and consumption. Yet instead of addressing the problem of overproduction, and the burden that it placed on the environment, government bureaucrats sought to address what they perceived as a problem of low consumption.

Federal support of marketing and transportation or direct government purchases for food relief or price control were all directed towards relieving industry depression by artificially manufacturing consumer demand. Temporarily increased demand largely disappeared after the passing of the immediate crisis that spurred it (war and postwar relief) or government efforts to create it (market manipulation). Thus, this consumer history of seafood exposes a fundamental flaw responsible for the historic underdevelopment of Canada's fishing industry. Throughout the twentieth century, the fisheries overproduced, seeking short-term profits through artificial market manipulation and creating long-term industry depression and eventually environmental collapse. In the end, the devastating boom-bust cycle was the product of not just fluctuations in fish supply but total miscalculation of consumer demand.

The efforts of the Department of Fisheries to manage consumer demand is part of the larger and longer history of Canada's National Policy. Beginning with the election of 1878, the federal government used its power to encourage western frontier settlement, build transportation and communication infrastructure, and create a domestic industrial economy via subsidies and tariffs. All these efforts, many historians have claimed, meant an active government promoting economic growth that concentrated power in central Canada. The fisheries were not exempt from the National Policy, despite what Atlantic Canadian historians have argued as part of larger discussions on the dependency of that region. More than ignoring the fisheries, the National Policy simply mismanaged them. Although the Department of Fisheries invested in scientific research, vessel subsidies, bounties, and other production-side stimuli, it also sought to aid the fisheries via consumption-side stimuli. By financing advertising and marketing to traditionally low-seafood-consuming regions and populations of Canada, industry and government leaders hoped to manufacture an enlarged domestic marketplace for Canada's fisheries products, thereby rescuing the historically underdeveloped industry from chronic depression while also partially liberating itself from its dependence on foreign markets.

These efforts might have succeeded as did similar other National Policy agendas, if industry and government leaders had listened to consumers and experts concerned with the low quality of Canada's seafood products. People like Evelene Spencer, Leonard Fraser, and the United Maritime Fishermen all expressed concerns with the industry and government's effort to market poor-quality fish, warning that such sales would have profound long-term consequences by discouraging potential consumers. The United Maritime Fishermen specifically targeted trawler fishing as the culprit for dumping low-grade fish on the marketplace. While their attack on steam trawlers was part of a larger political agenda, we should not disregard their firsthand observations of the poor quality of trawler-caught fish. In the end, however, industry and government leaders spurned inspection and regulation, hoping merely to craft a reliable supply of fish products to answer an assumed increase in consumer demand.

This consumer-oriented policy began during the first two decades of the twentieth century but became central to industry and government strategy by the 1920s and '30s. The agenda benefited from general trends of the early twentieth century that saw increased focus on marketing packaged, processed, and name-brand food products. Like other food producers, in their marketing the fisheries targeted nutrition and women. Both industry and government advertising projected a message that seafood fit within the new modern, middle-class diet focused on providing "wholesome" family meals. The discovery of vitamins and the mass distribution of that information through women's magazines, as well as industry food advertisements that pitched the nutritious qualities of their products, revolutionized the way consumers viewed the role of food in their health. The food industry, academia, and government all put pressure on women to make informed choices about the health qualities of the food they bought and prepared for their families. Home economics became the medium through which many women learned about the science of efficient, healthy mothering. Individual health and family health were linked to national (or public) health, and a woman's ability to provide healthy food for her family became a major part of a national solution to an assumed public health crisis of widespread malnourishment.

Government advertising of seafood, however, went beyond these established norms of the food industry. In addition to projecting the ideal of women solving the nation's health crisis, seafood advertising called upon women to revitalize one of the nation's most important and historic industries through their purchasing power. Marketing messages most often emphasized nutrition, ease of use, and the modernity of seafood as part of a middle-class diet; meanwhile, internal communications within the government and between government and industry regularly expressed expectations that increased purchasing by the nation's women would rescue the collapsing fisheries industry.

This focus on industry recovery, stability, and growth became even more central during the Great Depression. After high-profile government investigations beginning in the 1920s into chronic poverty in the industry, particularly in Atlantic Canada, the country became more aware of the economic realities of the nation's fishing communities. World war provided a stimulus for fisheries recovery, and Canada used its position in international organizations to facilitate large government purchases of seafood. In some sectors, government purchases for war or war relief consumed the majority, even the totality, of some of Canada's seafood products. This temporary surge in demand created a context in which few industry or government actors felt the need to examine more deeply the flaws of production and the perpetual imbalance between consumption and production.

By the early 1950s, there was finally a re-evaluation of the consumer-oriented strategy for fisheries recovery, stability, and growth. The Department of Fisheries at last concluded that increased production and marketing of seafood was not the solution. It switched instead to a focus on quality of production and began the regular inspection and regulation of seafood at all stages. Thus, this study comes to end where it began: with Evelene Spencer, hired by the department to educate the nation's homemakers on the value of seafood as part of a modern diet. Spencer's assigned mission symbolized the government's and the industry's ill-informed belief that the chronic depression of the nation's fisheries resulted from Canadian women's failure to embrace the nutritious and economic values of seafood and prepare it for their families in a pleasing way. But Spencer was too close to the front line of consumption to fully buy into this premise. Travelling the nation, lecturing and interacting with wholesalers, retailers, and customers, she perhaps knew better than anyone the real problem undermining seafood consumption. She only once hinted at what this study concludes – that Canadians did not buy more seafood because Canadian seafood was pretty bad. If industry and government wanted to use consumption as the means to restore economic value to the fisheries, they needed to improve quality. Instead, they dumped an increasing quantity of bad product on an already glutted market that was unlikely to want to buy more seafood. Under the false hope that more marketing would lead to more customers, the industry and government sought increased production. In the end, they only produced more fish than the market was able to absorb and ultimately more than Canada's oceans, lakes, rivers, and bays could sustainably provide.

Notes

INTRODUCTION

1 This controversy between inshore fishermen and trawlers, the "trawler question," dealt specifically with the development of steam-powered trawlers (hereafter referred to as either trawlers or steam trawlers). Trawling is a fishing technique in which a vessel drags a large, bag-like net through the fishing grounds. Although trawling dates back to the seventeenth century, the use of steam power first appeared in Britain's North Sea fishing fleet in the 1870s and came to dominate that country's fishing operations by the twentieth century. The first steam trawler in regular operation in Canada, the *Wren*, began operating out of Casco in 1908. Opposition to the trawler in Canada was fierce and led that same year to a government ban on trawling within three miles of the coast. In 1909, an Order-in-Council concluded that trawling was destructive to the fishing grounds and called for an international agreement banning offshore trawling. Although Canada, Newfoundland, and the United States all conducted investigations into the effect of trawling in 1912, no agreement materialized. In 1915, via customs regulation, Canada was effectively able to ban trawling within twelve miles of the coast, and thus the use of trawlers in Canada was significantly less than in the United States or United Kingdom. For a review of the issue, see H. Scott Gordon, "The Trawler Question in the United Kingdom and Canada," *Dalhousie Review* 31, no. 2 (1951): 117–27. Gordon was a Canadian economist best known for his "tragedy of the commons" thesis, first expressed in "Economic Theory of a Common Property Resource: The Fishery," in the *Journal of Political Economy* 62, no. 124 (1954). At that time, he taught at Carleton University in Ottawa and had also

worked for the Fisheries Price Support Board of the Department of Fisheries.

2 Harold Innis, *The Cod Fisheries: The History of an International Economy* (New Haven: Yale University Press, 1940).

3 Arthur F. McEvoy, *The Fisherman's Problem: Ecology and Law in California Fisheries, 1850–1980* (New York: Cambridge University Press, 1986), 251, 13–14.

4 Carmel Finley, *All the Boats on the Ocean: How Government Subsidies Led to Global Overfishing* (Chicago: University of Chicago Press, 2017), 2.

5 Ibid., 161–2.

6 Ibid., 51, 136, 145–6, 152. Maximum Sustainable Yield was based on the belief that scientists could calculate the amount of any resource that could be extracted without inflicting long-term damage. They almost always over-estimated that amount.

7 Paul Josephson, "The Ocean's Hot Dog: The Development of the Fish Stick," *Technology and Culture* 49, no. 1 (2008): 41–61. See also Kevin Bailey, *Billion-Dollar Fish: The Untold Story of Alaska Pollock* (Chicago: University of Chicago Press, 2013), 13, 39, 64.

8 Finley, *All the Boats on the Ocean*, 96, 131.

9 Great Britain, Sea Fisheries Commission, *Report of the Commissioners Appointed to Inquire into the Sea Fisheries of the United Kingdom*, vol. 1, "The Report and Appendix" (London: George Edward Eyre and William Spottiswoode, 1866), as specifically mentioned on pages xv–xx but continuously referred to throughout the study.

10 Some of the few fisheries histories that look at seafood include Carmel Finley's two books, *All the Fish in the Sea* and *All the Boats on the Ocean*, as well as Brian M. Fargan's *Fish on Friday: Feasting, Fasting, and the Discovery of the New World* (New York: Basic Books, 2006). Some local histories include Edmond Boudreaux, *The Seafood Capital of the World: Biloxi's Maritime History* (Gloucestershire: History Press, 2011), and Deanne Love Stephens, *The Mississippi Gulf Coast Seafood Industry: A People's History* (Jackson: University Press of Mississippi, 2021). There are also a host of "foodie" histories, often as much recipe books as history books. These include Anthony D. Fredericks, *The Secret Life of Clams: The Mystery and Magic of Our Favorite Shellfish* (New York: Skyhorse Publishing, 2014); Trevor Corson, *The Secret Life of Lobsters: How Fishermen and Scientists Are Unraveling the Mysteries of Our Favorite Crustacean* (New York: Harper Perennial, 2005); and Trevor Corson, *The Story of Sushi: An Unlikely Saga of Raw Fish and Rice* (New York: Harper Perennial, 2008).

11 Some of the better commodities histories include Elizabeth Abbott, *Sugar: A Bittersweet History* (Toronto: Penguin, 2008); Mark Kurlansky, *Salt: A World History* (New York: Penguin, 2003); Mark Kurlansky, *Birdseye: The Adventures of a Curious Man* (New York: Doubleday, 2012); Sidney W. Mintz, *Sweetness and Power: The Place of Sugar in Modern History* (New York: Penguin, 1985); Mark Pendergrast, *Uncommon Grounds: The History of Coffee and How It Transformed Our World* (New York: Basic Books, 1999); John Reader, *Potato: A History of the Propitious Esculent* (New Haven, CT: Yale University Press, 2008); Daniel Robinson, "Marketing Gum, Making Meaning: Wrigley in North America, 1890–1930," *Enterprise and Society* 5, no. 1 (2004): 4–44; Radcliffe N. Salaman, *The History and Social Influence of the Potato* (Cambridge, MA: Cambridge University Press, 1985); Daniel Sidorick, *Condensed Capitalism: Campbell's Soup Company and the Pursuit of Cheap Production in the Twentieth Century* (Ithaca: Cornell University Press, 2009); Andrew F. Smith, *Popped History: A Social History of Popcorn in America* (Columbia: University of South Carolina Press, 1999); John Soluri, *Banana Cultures: Agriculture, Consumption, and Environmental Change in Honduras and the United States* (Austin: University of Texas Press, 2005); Carolyn Wyman, *Spam: A Biography* (San Diego: Harcourt Brace, 1999). For ethnic- and gender-based studies of food, see Arlene Voski Avakian and Barbara Haber, eds., *From Betty Crocker to Feminist Food Studies: Critical Perspectives on Women and Food* (Amherst: University of Massachusetts Press, 2005); Franca Iacovetta and Valerie J. Korinek, "Jell-O Salad, One-Stop Shopping, and Maria the Homemaker: The Gender Politics of Food," in *Sisters or Strangers? Immigrant, Ethnic, and Radicalized Women in Canadian History*, ed. Marlene Epp, Franca Iacovetta, and Frances Swyria (Toronto: University of Toronto Press, 2004); Sherrie A. Inness, *Dinner Roles: American Women and Culinary Culture* (Iowa City: University of Iowa Press, 2001); Sherrie A. Inness, ed., *Cooking Lessons: The Politics of Gender and Food* (Lanham, MD: Rowman and Littlefield, 2001); Sherrie A. Inness, *Secret Ingredients: Race, Gender, and Class at the Dinner Table* (London: Palgrave Macmillan, 2006); Laura Shapiro, *Perfection Salad: Women and Cooking at the Turn of the Century* (New York: Farrar, Straus and Giroux, 1986); Janet Theophano, *Eat My Words: Reading Women's Lives through the Cookbooks They Wrote* (New York: Palgrave, 2002); Psyche A. Williams-Forson, *Building Houses out of Chicken Legs: Black Women, Food, and Power* (Chapel Hill: University of North Carolina Press, 2006). For food studies that address food as a social equity issue or political tool, see Rima Apple, *Vitamania: Vitamins in American Cul-*

ture (New Brunswick, NJ: Rutgers University Press, 1996); Nick Callather, "The Foreign Policy of Calories," *American Historical Review* 112, no. 2 (2007): 1–60; John Coveney, *Food Morals and Meaning: The Pleasure and Anxiety of Eating*, 2nd ed. (London: Routledge, 2007); Hasia R. Diner, *Hungering for America: Italian, Irish, and Jewish Foodways in the Age of Migration* (Cambridge, MA: Harvard University Press, 2001); Harmke Kamminga and Andrew Cunningham, eds., *Science and Culture of Nutrition* (Amsterdam: Rodophi, 1995); Harvey Levenstein, *Revolution at the Table: The Transformation of the American Diet* (New York: Oxford University Press, 1988); Harvey Levenstein, *Paradox of Plenty: A Social History of Eating in Modern America* (New York: Oxford University Press, 1993); David F. Smith, ed., *Nutrition in Britain: Science, Scientists, and Politics in the Twentieth Century* (London: Routledge, 1997); David F. Smith and Jim Philips, eds., *Food, Science, Policy, and Regulation in the Twentieth Century: International and Comparative Perspectives* (London: Routledge, 2000); Jane Ziegelman, *A Square Meal: A Culinary History of the Great Depression* (New York: Harper, 2016). For Canadian food histories, see W.H. Heick, *A Propensity to Protect: Butter, Margarine and the Rise of Urban Culture in Canada* (Waterloo, ON: Wilfred Laurier University Press, 1991); Hermiston, "'If It's Good for You, It's Good for the Nation!'"; Franca Iacovetta, Valerie J. Korinek, and Marlene Epp, eds., *Edible Histories, Cultural Politics: Towards a Canadian Food History* (Toronto: University of Toronto Press, 2012); Douglas McCalla, "The Wheat Staple in Upper Canadian Development," *Canadian Historical Association Historical Papers*, no. 13 (1978): 34–6; John McCallum, *Unequal Beginnings: Agriculture and Economic Development in Ontario and Quebec* (Toronto: University of Toronto Press, 1980); Ian Mosby, *Food Will Win the War: The Politics, Culture, and Science of Food on Canada's Home Front* (Vancouver: University of British Columbia Press, 2014); Alex Ostry, "The Interplay of Public Health and Economics in the Early Development of Nutrition Policy in Canada," *Critical Public Health* 13, no. 2 (2003): 171–85; Alex Ostry, *Nutrition Policy in Canada, 1870–1939* (Vancouver: University of British Columbia Press, 2006); John F. Varty, "On Protein, Prairie Wheat and Good Bread: Rationalizing Technologies and the Canadian State, 1912–1935," *Canadian Historical Review* 85, no. 4 (2004): 721–53.

12 Popular books that approach the fisheries story from the perspective of food studies include Bailey, *Billion-Dollar Fish;* Paul Greenberg, *Four Fish: The Future of the Last Wild Food* (New York: Penguin Books, 2011); Paul Greenberg, *American Catch: The Fight for Our Local Seafood* (New

York: Penguin Books, 2014); David G. Gordon, Nancy E. Blanton, and Terry Y. Nosho, *Heaven on the Half Shell: The Story of the Northwest's Love Affair with the Oyster* (Portland, OR: WestWinds Press, 2003); Drew Smith, *Oyster: A Gastronomic History (with Recipes)* (New York: Harry N. Abrams, 2015).

13 Psalms 107:23–31 (King James Version).

14 Jean-François Bière, "The French Fishery in North America in the 18th Century," in *How Deep Is the Ocean? Historical Essays on Canada's Atlantic Fishery*, ed. James Candow and Carol Corbin (Sydney: University College of Cape Breton Press, 1997); Sean Cadigan, *Hope and Deception in Conception Bay: Merchant-Settler Relations in Newfoundland, 1785–1855* (Toronto: University of Toronto Press, 1995); Sean Cadigan, "Failed Proposals for Fisheries Management and Conservation in Newfoundland, 1855–1880," in *Fishing Places, Fishing People: Issues in Small Scale Fisheries*, ed. Dianne Newell and Rosemary Ommer (Toronto: University of Toronto Press, 1998); Sean Cadigan, "The Moral Economy of the Commons: Ecology and the Equity in Newfoundland Cod Fisheries, 1815–1855," *Labour/Le Travail* 43 (1999): 9–42; John E. Crowley, "Empire versus Trunk: The Official Interpretation of Debt and Labour in the Eighteenth-Century Newfoundland Fishery," *Canadian Historical Review* 70, no. 3 (1989): 311–36; Jennifer Hubbard, *A Science on the Scale: The Rise of Canadian Fisheries Biology, 1898–1939* (Toronto: University of Toronto Press, 2006); David A. MacDonald, "They Cannot Pay Us in Money: Newman and Company and the Supplying System in the Newfoundland Fishery, 1850–1884," *Acadiensis* 19, no. 1 (1989): 142–56; Dianne Newell and Rosemary Ommer, *Fishing Places, Fishing People: Traditions and Issues in Canadian Small-Scale Fisheries* (Toronto: University of Toronto Press, 1999); Rosemary Ommer, *From Outpost to Outpost: A Structural Analysis of the Jersey-Gaspe Cod Fishery, 1767–1886* (Montreal and Kingston: McGill-Queen's University Press, 1991); Brian Payne, *Fishing a Borderless Sea: Environmental Territorialism in the North Atlantic, 1818–1910* (East Lansing: Michigan State University Press, 2010); Brian Payne, "Becoming a Dependent Class: Quoddy Herring Fishermen in the 1920s," *Labour/Le Travail* 81 (2018): 87–117; Peter Pope, *Fish into Wine: The Newfoundland Plantation in the Seventeenth Century* (Chapel Hill: University of North Carolina Press, 2004); Mariam Wright, *A Fishery for Modern Times: The State and Industrialization of the Newfoundland Fisheries, 1934–1968* (New York: Oxford University Press, 2001).

15 W. Jeffrey Bolster, *The Mortal Sea: Fishing the Atlantic in the Age of Sail* (Cambridge: Harvard University Press, 2014); Michael J. Chiarappa,

"Dockside Landings and Threshold Spaces: Reckoning Architecture's Place in Marine Environmental History," *Environmental History* 18, no. 1 (2013): 12–28; Richard Judd, "Grass-Roots Conservation in Eastern Coastal Maine: Monopoly and the Moral Economy of Weir Fishing, 1893–1911," *Environmental Review* 12, no. 2 (1988): 81–103; Richard Judd, "Saving the Fishermen as Well as the Fish: Conservation and Commercial Rivalry in Maine's Lobster Industry: 1872–1933," *Business History Review* 62, no. 4 (1988): 596–625; Matthew McKenzie, *Clearing the Coastline: The Nineteenth-Century Ecological and Cultural Transformation of Cape Cod* (Lebanon, NH: University of New England Press, 2011); Matthew McKenzie, *Breaking the Banks: Representations and Realities in New England Fisheries, 1866–1966* (Amherst: University of Massachusetts Press, 2018); Wayne O'Leary, *Maine Sea Fisheries: The Rise and Fall of a Native Industry* (Boston: Northeastern University, 1996); Daniel Vickers, *Farmers and Fishermen: Two Centuries of Work in Essex County, Massachusetts, 1630–1850* (Chapel Hill: University of North Carolina Press, 1994).

16 Lisa A. Brown, *The Nature of Borders: Salmon, Boundaries and Bandits on the Salish Sea* (Seattle: University of Washington Press, 2012); Connie Y. Chiang, *Shaping the Shoreline: Fisheries and Tourism on the Monterey Coast* (Seattle: University of Washington Press, 2001); Matthew D. Evenden, *Fish versus Power: An Environmental History of the Fraser River* (New York: Cambridge University Press, 2004); Arthur F. McEvory, *The Fisherman's Problem: Ecology and Law in California Fisheries, 1850–1980* (New York: Cambridge University Press, 1986); Joseph Taylor, *Making Salmon: An Environmental History of the Northwest Fisheries Crisis* (Seattle: University of Washington Press, 2001).

17 Finley, *All the Fish in the Sea*; Finley, *All the Boats on the Ocean*.

18 Finley, *All the Fish in the Sea*, 8.

19 Ibid., 27.

20 Ibid., 23.

21 Ibid., 40.

22 Some colonial American fisheries histories do discuss Native American fishing culture and early colonial use of fish for subsistence. See Matthew McKenzie, *Clearing the Coastline*; Erik Reardon, *Managing the River Commons: Fishing and New England's Rural Economy* (Amherst: University of Massachusetts Press, 2021). See also Eric Mills, *Biological Oceanography: An Early History, 1870–1960* (Toronto: University of Toronto Press, 2012). There has been some discussion of this topic in Norwegian and Sami culture as well. See Lars Ivar Hansen, "Sami Fishing in the Pre-Modern Era: Household Sustenance and Market Rela-

tions" 65–82, and Jòn Th. Thór, "Icelandic Fisheries c. 900–1900," 323–49, both in *A History of the North Atlantic* Fisheries, vol. 1, *From Early Times to the Mid-Nineteenth Century*, ed. David J. Starkey, Jòn Th. Thór, and Ingo Heidbrink (Bremen: Verlag H.M. Hauschild, 2009).

23 Canadian involvement in international cooperation in fisheries conservation began in the 1920s with its work with and involvement in the North American Council on Fisheries Investigations (NACFI).

CHAPTER ONE

1 John L. Finlay and D.N. Sprague, *The Structure of Canadian History*, 3rd ed. (Scarborough, ON: Prentice-Hall Canada, 1989), 215.

2 Charles Perry Stacey, *Canada and the Age of Conflict*, vol. 1, *1867–1921* (Toronto: University of Toronto Press, 1984), 32.

3 I discuss the impact of tariffs, reciprocity with the United States, and transportation links as they specifically relate to Prince Edward Island in the mid-nineteenth century in "'The Best Fishing Station': The Fish Trade of Prince Edward Island and Resource Transfer in the Gulf of St. Lawrence, 1854–1873," in *The Greater Gulf: Essays on the Environmental History of the Gulf of St. Lawrence*, ed. Claire Campbell, Edward Mac-Donald, and Brian Payne (Montreal and Kingston: McGill-Queen's University Press, 2019).

4 Finlay and Sprague, *Structure of Canadian History*, 234.

5 Stacey, *Canada and the Age of Conflict*, 37.

6 Ibid., 38–40.

7 As quoted in ibid., 89.

8 Ibid., 85–90.

9 Ibid., 102.

10 Ibid., 143–9.

11 Finlay and Sprague, *Structure of Canadian History*, 275.

12 Matthias Blum, "War, Food, Rationing, and Socioeconomic Inequality in Germany during the First World War," *Economic History Review* 66, no. 4 (November 2013): 1063–83; Mary Elisabeth-Cox, "Hunger Games: Or How the Allied Blockade in the First World War Deprived German Children of Nutrition, and Allied Food Aid Subsequently Saved Them," *Economic History Review* 68, no. 2 (2015): 600–31; David C. Hsiung, "Food, Fuel, and the New England Environment in the War for Independence, 1775–1776," *New England Quarterly* 80, no. 4 (2007): 614–55; Erik-C. Landis, "Between Village and Kremlin: Confronting State Food Procurement in Civil War Tambov, 1919–20," *Russian Review* 63, no. 1

(2004): 70–88; Gregory H. Maddox, "Njaa: Food Shortages and Famines in Tanzania between the Wars," *International Journal of African Historical Studies* 19, no. 1 (1986): 17–34; Anthony Oberschall and Michael Seidman, "Food Coercion in Revolution and Civil War: Who Wins and How They Do It," *Comparative Studies in Society and History* 47, no. 2 (2005): 372–402; Melanie Schulze Tanielian, "Feeding the City: The Beirut Municipality and the Politics of Food during World War I," *International Journal of Middle East Studies* 46, no. 4 (2014): 737–58.

13 William Knight, "Modeling Authority at the Canadian Fisheries Museum, 1884–1918," PhD diss., Carleton University, 2014, 251–96.

14 "Back Up the Troops by Substituting," *Canada Food Bulletin* (Ottawa), Thursday, 28 March 1918, no. 13, 24; "Subdue the Submarine," *Canada Food Bulletin* (Ottawa), Saturday, 9 February 1918, no. 10, 11.

15 Such advertisements can be found in "Report of Canada Food Board," Ottawa, 13 February 1919, General Information, Correspondence Concerning Encouragement of Production of Fish to Help Relieve Shortage of Food (hereafter Correspondences Re: Production of Fish), 31 January 1919 to 22 November 1919, Library and Archives Canada (hereafter LAC), RG 23, vol. 509, folder 711-1-33. See also "Appendix to the Report of the Food Controller, Hon. W.J. Hanna, K.C., 1918, Fish," General Information, Expansion of the Fishing Trade, Survey of Fishing Industry, 26 March 1931 to 4 January 1932, LAC, RG 23, vol. 531, folder 711-25-22 (2).

16 A Statement and Appeal, Ottawa, 7 August 1917, General Information, Correspondences Re: Production of Fish, 31 January 1919 to 22 November 1919, LAC, RG 23, vol. 509, folder 711-1-33.

17 Ibid.

18 These Canada Food Board advertisements are located in General Information, Expansion of Fishing Trade, Value of Train-Ferries, LAC, RG 23, vol. 527, folder 711-25-4.

19 "Duty Says: Eat Fish!," advertisement for John Brown & Co., Purveyors, *Canadian Grocer*, 26 October 1917, 116.

20 "Eat and Enjoy," advertisement for Connors Bros. Ltd, Black Harbor, NB, *Canadian Grocer*, 2 November 1917, 54.

21 "Eat Fish and Vegetables: These Are the Only Substitutes Now Available in Large Quantities," *Canada Food Bulletin* (Ottawa), Saturday, 1 December 1917, no. 5, 17.

22 "Fish and Chips: They Helped Immeasurably in the Days of Food Shortage in England – They Furnish a Delicious and Nourishing Meal for Fifteen Cents," n.d., Canada Food Board Promotion Division, for

Release to Evening Papers, General Information, Correspondences Re: Production of Fish, LAC, RG 23, vol. 509, folder 711-1-22 (3).

23 "Good Deep Sea Fish Plentiful and Cheap," *Daily Colonist* (Victoria, BC), May 1918, included in General Information, Correspondences Re: Production of Fish, LAC, RG 23, vol. 509, folder 711-1-22 (3).

24 "Fish and Chips: They Helped Immeasurably in the Days of Food Shortage [...]," n.d., Canada Food Board Promotion Division, for Release to Evening Papers, General Information, Correspondences Re: Production of Fish, LAC, RG 23, vol. 509, folder 711-1-22 (3).

25 The cook-booklet and advertising can be found in General Information, Correspondences Re: Production of Fish, LAC, RG 23, vol. 509, folder 711-1-22 (3).

26 "An Appeal for Increased Production of Fish," Deputy Minister of the Naval Service, Ottawa, 6 August 1917, General Information, Correspondences Re: Production of Fish, 1 July 1917 to 18 September 1917, LAC, RG 23, vol. 509, folder 711-1-33 (1).

27 Advertisement for "An Interesting Film of Atlantic Fishing for Cod, Haddock, Pollock, Herring, and Other Fish," 13 February 1919, General Information, Correspondences Re: Production of Fish, LAC, RG 23, vol. 509, folder 711-1-22 (3).

28 Letters from W. Fisher, Asst. Superintendent of Fisheries, 20 July 1917, sent to regional fisheries inspectors in Atlantic (Matheson, Marshall, Hockin, McLeod, Harrison, Morrison, Calder, and Bernier), 31 January 1919 to 22 November 1919, LAC, RG 23, vol. 509, folder 711-1-33.

29 G. Frank Beer, Esq., R.Y. Eaton, Esq, Fish Committee, Office of the Food Controller, Ottawa, to C.C. Ballantyne, Minister of Marine and Fisheries, Ottawa, 14 November 1917, Taking over Fishery Service, Correspondence Concerning Proposal of Food Controller for the Placing of Fisheries Branch Under Fish Committee, November 1917, LAC, RG 23, vol. 525, folder 711-5-5.

30 C.C. Ballantyne, Minister of Marine and Fisheries, Ottawa, to G. Frank Beer, Esq., R.Y. Eaton, Esq., Fish Committee, 28 November 1917, Taking over Fishery Service [...], November 1917, LAC, RG 23, vol. 525, folder 711-5-5.

31 Frederick Wallace, Memorandum on the Fish Section for the Report of Canada Food Board, Ottawa, 13 February 1919, 31 January 1919 to 22 November 1919, LAC, RG 23, vol. 509, folder 711-1-33.

32 G.J. Desbarats, Deputy Minister of Naval Service, to H.B. Thomson, Esq., Chairman, Canada Food Board, 6 March 1919, 31 January 1919 to 22 November 1919, LAC, RG 23, vol. 509, folder 711-1-33.

33 "The Why in Meat Eating," Canada Food Board, 1919, General Information, Correspondences Re: Production of Fish, LAC, RG 23, vol. 509, folder 711-1-22 (3).

34 "Fish Alive-O! The Diet of Health Canada's New Wealth: Why Every Canadian Housewife Should Learn to Cook Fish," Canada Food Board, Ottawa, August 1918, General Information, Correspondences Re: Production of Fish, LAC, RG 23, vol. 509, folder 711-1-22 (3).

35 "Apply Thrift and Live Better: The Need for Care While Canada Has to Face After-the-War Problems," Canada Food Board, 1919, General Information, Correspondences Re: Production of Fish, LAC, RG 23, vol. 509, folder 711-1-22 (3); "Eat Fish and Vegetables: These Are the Only Substitutes Now Available in Large Quantities," *Canada Food Bulletin*, Saturday, 1 December 1917, 17.

36 "Canada's Fish Education," advertisement for Connors Brothers Ltd, *Canadian Grocer*, 21 September 1917, 54.

37 Finlay and Sprague, *Structure of Canadian History*, 248.

38 Ibid., 251.

39 David Alexander, "Economic Growth in the Atlantic Region, 1880–1940," in *The Acadiensis Reader*, vol. 2, *Atlantic Canada after Confederation*, ed. P.A. Buckner and David Frank, 2nd ed. (Fredericton, NB: Acadiensis Press, 1988), 135.

40 Ibid., 135.

41 Ibid., 154–5.

42 Patricia A. Thornton, "The Problem of Out-Migration from Atlantic Canada, 1871–1921: A New Look," in Buckner and Frank, *Acadiensis Reader*, 2:34–65.

43 T.W. Acheson, "The National Policy and the Industrialization of the Maritimes, 1880–1900," in Buckner and Frank, *Acadiensis Reader*, 2:164–89.

44 Finlay and Sprague, *Structure of Canadian History*, 301; E.R. Forbes, "The Origins of the Maritime Rights Movement," *Acadensis* 5, no. 1 (1975): 55–61.

45 Acheson, "National Policy and the Industrialization of the Maritimes," 189.

46 Jas. H. Colon, Director, Publicity, Marketing, and Transportation Division, to F.W. Wallace, Secretary, Canadian Fisheries Ass'n., Gardenvale, Quebec, 4 January 1921, General Information, Correspondences Re: Production of Fish, LAC, RG 23, vol. 509, folder 711-1-22 (5).

47 Memo from G.C. Ballantyne, Minister of Marine and Fisheries, 21 January 1921, General Information, Correspondence Regarding Encouraging the Production of Fish to Help the Shortage of Food, LAC, RG 23, vol. 509, folder 711-1-22 (5).

48 "Fishermen of the Breed That Builds Empires," National Fish Day Propaganda Article no. 2, Department of Marine and Fisheries, mailed 7 January 1922, Department of Marine and Fisheries, National Fish Day, General Correspondences, January 1922 to 6 February 1922, LAC, RG 23, vol. 536, folder 711-1-33 (6).

49 E. Hawken, Acting Deputy Minister of Marine and Fisheries, to Dr J.H. Putman, Inspector of Schools, Ottawa, 24 January 1921, General Information, Correspondences Re: Production of Fish, 20 November 1919 to 29 December 1921, LAC, RG 23, vol. 509, folder 711-1-22 (5).

50 Memorandum from W. Found, Assistant Deputy Minister of Fisheries, Department of Marine and Fisheries, n.d., General Information, Correspondences Re: Production of Fish, 20 November 1919 to 29 December 1921, LAC, RG 23, vol. 509, folder 711-1-22 (5).

CHAPTER TWO

1 Some of the better commodities history include Elizabeth Abbott, *Sugar: A Bittersweet History* (Toronto: Penguin, 2008); Mark Kurlansky, *Salt: A World History* (New York: Penguin, 2003); Mark Kurlansky, *Birdseye: The Adventures of a Curious Man* (New York: Doubleday, 2012) Sidney W. Mintz, *Sweetness and Power: The Place of Sugar in Modern History* (New York: Penguin, 1985); Mark Pendergrast, *Uncommon Grounds: The History of Coffee and How It Transformed Our World* (New York: Basic Books, 1999); John Reader, *Potato: A History of the Propitious Esculent* (New Haven: Yale University Press, 2008); Daniel Robinson, "Marketing Gum, Making Meaning: Wrigley in North America, 1890–1930," *Enterprise and Society* 5, no. 1 (2004): 4–44; Radcliffe N. Salaman, *The History and Social Influence of the Potato* (Cambridge, MA: Cambridge University Press, 1985); Daniel Sidorick, *Condensed Capitalism: Campbell's Soup Company and the Pursuit of Cheap Production in the Twentieth Century* (Ithaca: Cornell University Press, 2009); Andrew F. Smith, Andrew, *Popped History: A Social History of Popcorn in America* (Columbia: University of South Carolina Press, 1999); John Soluri, *Banana Cultures: Agriculture, Consumption, and Environmental Change in Honduras and the United States* (Austin: University of Texas Press, 2005); Carolyn Wyman, *Spam: A Biography* (San Diego: Harcourt Brace, 1999).

2 See, for example, Arlene Voski Avakian and Barbara Haber, eds., *From Betty Crocker to Feminist Food Studies: Critical Perspectives on Women and Food* (Amherst: University of Massachusetts Press, 2005); Franca Iacovetta and Valerie J. Korinek, "Jell-O Salad, One-Stop Shopping, and Maria

the Homemaker: The Gender Politics of Food," in *Sisters or Strangers? Immigrant, Ethnic, and Radicalized Women in Canadian History*, ed. Marlene Epp, Franca Iacovetta, and Frances Swyria (Toronto: University of Toronto Press, 2004); Sherrie A. Inness, *Dinner Roles: American Women and Culinary Culture* (Iowa City: University of Iowa Press, 2001); Sherrie A. Inness, ed., *Cooking Lessons: The Politics of Gender and Food* (Lanham, MD: Rowman & Littlefield, 2001); Sherrie A. Inness, *Secret Ingredients: Race, Gender, and Class at the Dinner Table* (London: Palgrave Macmillan, 2006); Laura Shapiro, *Perfection Salad: Women and Cooking at the Turn of the Century* (New York: Farrar, Straus and Giroux, 1986); Janet Theophano, *Eat My Words: Reading Women's Lives through the Cookbooks They Wrote* (New York: Palgrave, 2002); Psyche A. Williams-Forson, *Building Houses out of Chicken Legs: Black Women, Food, and Power* (Chapel Hill: University of North Carolina Press, 2006).

3 Rima Apple, *Vitamania: Vitamins in American Culture* (New Brunswick, NJ: Rutgers University Press, 1996); Nick Callather, "The Foreign Policy of Calories," *American Historical Review* 112, no. 2 (2007): 1–60; John Coveney, *Food Morals and Meaning: The Pleasure and Anxiety of Eating*, 2nd ed. (London: Routledge, 2007); Hasia R. Diner, *Hungering for America: Italian, Irish, and Jewish Foodways in the Age of Migration* (Cambridge: Harvard University Press, 2001); Harmke Kamminga and Andrew Cunningham, eds., *Science and Culture of Nutrition* (Amsterdam: Rodophi, 1995); Harvey Levenstein, *Revolution at the Table: The Transformation of the American Diet* (New York: Oxford University Press, 1988); Harvey Levenstein, *Paradox of Plenty: A Social History of Eating in Modern America* (New York: Oxford University Press, 1993); David F. Smith, ed., *Nutrition in Britain: Science, Scientists, and Politics in the Twentieth Century* (London: Routledge, 1997); David F. Smith and Jim Philips, eds., *Food, Science, Policy, and Regulation in the Twentieth Century: International and Comparative Perspectives* (London: Routledge, 2000); Jane Ziegelman, *A Square Meal: A Culinary History of the Great Depression* (New York: Harper, 2016).

4 W.H. Heick, *A Propensity to Protect: Butter, Margarine and the Rise of Urban Culture in Canada* (Waterloo, ON: Wilfred Laurier University Press, 1991); Alana J. Hermiston, "'If It's Good for You, It's Good for the Nation!': The Moral Regulation of Nutrition in Canada, 1930–1945" (PhD diss., Carleton University, 2005); Franca Iacovetta, Valerie J. Korinek, and Marlene Epp, eds., *Edible Histories, Cultural Politics: Towards a Canadian Food History* (Toronto: University of Toronto Press, 2012); Douglas McCalla, "The Wheat Staple in Upper Canadian Development,"

Canadian Historical Association Historical Papers (1978): 34–6; John McCallum, *Unequal Beginnings: Agriculture and Economic Development in Ontario and Quebec* (Toronto: University of Toronto Press, 1980); Ian Mosby, *Food Will Win the War: The Politics, Culture, and Science of Food on Canada's Home Front* (Vancouver: University of British Columbia Press, 2014); Alex Ostry, "The Interplay of Public Health and Economics in the Early Development of Nutrition Policy in Canada," *Critical Public Health* 13, no. 2 (2003): 171–85; Alex Ostry, *Nutrition Policy in Canada, 1870–1939*. (Vancouver: University of British Columbia Press, 2006); John F. Varty, "On Protein, Prairie Wheat and Good Bread: Rationalizing Technologies and the Canadian State, 1912–1935," *Canadian Historical Review* 85, no. 4 (2004): 721–53.

5 Apple, *Vitamania*; Catherine Price, *Vitamania: Our Obsessive Quest for Nutritional Perfection* (New York: Penguin, 2015). For a contemporary understanding of the process of nutritional science and the knowledge of vitamins, and their impact on flour and bread, see Russell M. Wilder, "Public Health Aspects of Enriched Flour and Bread," paper presented at Millers Conference in Chicago, 5 March 1941, and printed by the Nutrition Advisory Committee to the Coordination of Health, Welfare, and Related Defense Activities, United States of America. Wilder was professor of medicine and chief of the Department of Medicine at the Mayo Clinic, chair of the Committee on Food Nutrition of the National Research Council, and a member of the Nutrition Advisory Committee to the Coordinator of Health, Welfare, and Related Defense Activities. Offprint of paper located in folder Dom. Cerealist – Vitamins in Bread, Wheat, and Flour, May 1941, LAC, RG 17, vol. 4038.

6 See particularly Coveney, *Food Morals and Meaning*; Ziegelman, *Square Meal*; also Callather, "Foreign Policy of Calories"; James, *Hunger.*

7 Mosby, *Food Will Win the War*, 36.

8 Ibid., 25–6, 30–1.

9 Ibid., 31–5.

10 "Fish and Cure Fish: To Close Canneries," *Canadian Grocer*, 4 March 1904, 30.

11 "Teach Your Customers the Fish Habit," advertisement for Black Bros. & Co., Ltd, in *Canadian Grocer*, 4 March 1904.

12 Numerous ads in the *Canadian Grocer*, especially those for name-brand canned salmon, emphasized the cleanliness of the factory. See, for example, "The Fanciest Quality of Red Sockeye Packed on the Fraser River," advertisement for Clover Leaf Salmon, of the Pacific Selling Co., *Canadian Grocer*, 8 July 1904, 47; "Are the Choicest in the Market," advertise-

ment for Sovereign and Lynx Brand Sockeye Salmon, *Canadian Grocer*, 8 July 1904, 25; "For 25 Years the Standard in Canada; Every Can Guaranteed; The Quality Unexcelled," advertisement for Horseshoe Salmon, *Canadian Grocer*, 15 July 1904, 33.

13 "A Sign of Good Quality," advertisement for Atlantic Fish Company Ltd, Lunenburg, *Canadian Grocer*, 19 October 1906, 21.

14 "The Quality Brand," advertisement for Connors Brothers Ltd, *Canadian Grocer*, 6 January 1911, 58; "Brunswick Brand Sea Food," advertisement for Connors Brothers Ltd, *Canadian Grocer*, 16 February 1912, 72.

15 "Brunswick Brand," advertisement for Connors Brothers Ltd, *Canadian Grocer*, 13 January 1911, 61.

16 "Ocean Brand," advertisement for Ocean Brand Seafood, *Canadian Grocer*, 13 January 1911, 59.

17 Ann Adam, "Fish – Any Time, Any Place," *Canadian Home Journal*, February 1932, 46, 50b.

18 "And Now the Key-Opening Brunswick," advertisement for Connors Brothers Ltd, *Canadian Grocer*, 25 November 1924, 6–7; "The New Way to Sell Sardine," advertisement for Connors Brothers Ltd, *Canadian Grocer*, 12 February 1926, 11.

19 "Unvarying Excellence," advertisement for Connors Brothers Ltd, *Canadian Grocer*, 27 October 1916, 30. See also "Quality Did It," advertisement for Connors Brothers Ltd, *Canadian Grocer*, 28 January 1916, 54; "Selling Brunswick Brand Sea Foods Will Add to Your Prestige," advertisement for Connors Brothers Ltd, *Canadian Grocer*, 24 March 1916, 8; "The Public Has Confidence in Brunswick Brand," advertisement for Connors Brothers Ltd, *Canadian Grocer*, 14 April 1916, 4; "This Is a Profit Maker," advertisement for Connors Brothers Ltd, *Canadian Grocer*, 12 May 1916, 58. By 1924, Connors Brothers was claiming that 75 per cent of all sardines sold in Canada were Brunswick Brand and that they sold three times as many as all other companies combined. See "The Demand for Sardines in Canada Has Only Begun!," advertisement for Connors Brothers Ltd, *Canadian Grocer*, 25 April 1924, 37; "75% of Sardine Sales in Canada Are Brunswick – 15 Million Tin Production!," advertisement for Connors Brothers Ltd, *Canadian Grocer*, 2 May 1924, 148–9; "Three out of Four Tins of Sardines Sold in Canada Are Packed by Connors Bros.," advertisement for Connors Brothers Ltd, *Canadian Grocer*, 12 September 1924, 81.

20 "The Best Fish in the World Is 'Halifax' Prepared Codfish," advertisement for Black Bros. & Co. Ltd, *Canadian Grocer*, 24 November 1905, 57.

21　"Sell Brunswick Brand Sea Foods for Profits and Reputation," advertisement for Connors Brothers Ltd, *Canadian Grocer*, 1 January 1915, 56; "The Select of the Fishermen's Catch," advertisement for Connors Brothers Ltd, *Canadian Grocer*, 8 January 1915, 8.

22　"This Lenten Season," advertisement for Connors Brothers Ltd, *Canadian Grocer*, 15 March 1915, 56.

23　"It's Knowledge That Directs the Successful Grocer," advertisement for Connors Brothers Ltd, *Canadian Grocer*, 19 March 1915, 56; "Canned Sea Foods of Established Quality," advertisement for Connors Brothers Ltd, *Canadian Grocer*, 16 April 1915, 45; "Keeping Up Sales," advertisement for Connors Brothers Ltd, *Canadian Grocer*, 27 April 1915, 56; "A Delightful Change from the Continual Breakfast of Bacon and Eggs," advertisement for Connors Brothers Ltd, *Canadian Grocer*, 16 June 1915, 50.

24　"Unvarying Excellence," advertisement for Connors Brothers Ltd, *Canadian Grocer*, 27 October 1916, 30. See also "There's a Quality Reputation Back of the Brunswick Brand Trade Mark," advertisement for Connors Brothers Ltd, *Canadian Grocer*, 6 April 1917, 46; "The Kind That Pleases the Discriminating," advertisement for Connors Brothers Ltd, *Canadian Grocer*, 4 May 1917, 44; "Eat and Enjoy," advertisement for Connors Brothers Ltd, *Canadian Grocer*, 2 November 1917, 54; "Fish Foods Selected, Prepared and Packed by Experts," advertisement for Connors Brothers Ltd, *Canadian Grocer*, 6 January 1922, 46.

25　"Talking about Fast Sellers – Did You Every Try These Popular 'Brunswick Brand' Sea Foods," advertisement for Connors Brothers Ltd, *Canadian Grocer*, 17 February 1922, 45. The company also emphasized the "confidence of Canadian women" in their product in "Why Are They Such Fast Sellers?," advertisement for Connors Brothers Ltd, *Canadian Grocer*, 17 March 1922, 45.

26　"Codfish Philosophy," *Canadian Grocer*, 21 October 1904, 115.

27　"Talk Fish to Your Customer," advertisement for Black Bros. & Co. Ltd, *Canadian Grocer*, 20 January 1904. Emphasis in original. The slogan "Better Than Meat" appeared often in reference to the health qualities of fish in Black Brothers' ads. See "A Healthy Purse," advertisement for Black Bros. & Co. Ltd, *Canadian Grocer*, 20 May 1904, 81.

28　"Teach Your Customers the Fish Habit," advertisement for Black Bros. & Co. Ltd, *Canadian Grocer*, 4 March 1904.

29　See, for example, "How to Handle Fish," 15 July 1904, 23; "Fish as Food," *Canadian Grocer*, 19 August 1904, 26; "To Benefit Canadian Fishermen," *Canadian Grocer*, 7 October 1904, 17; "Source of Our Fish Supply," *Cana-*

dian Grocer, 21 October 1904, 115; "Interview with Montreal Fish Dealer," *Canadian Grocer*, 17 March 1905, 19; "Experience Teaches the Grocer That It's a Good Thing to Push Fish," *Canadian Grocer*, 19 May 1905, 16; E.A. Hughes, "Money in Selling Fish," *Canadian Grocer*, 29 October 1915, 115–16; "Merchants' Experience in Handling Fish," *Canadian Grocer*, 26 October 1917, 119, 121; "Eaton's Installs Elaborate Fish Dept.," *Canadian Grocer*, 26 October 1917, 120; T.M. Fraser, "Keeping after Fish Sales," *Canadian Grocer*, 3 October 1919, 1; "'Cleanliness' the Fish Man's Slogan," *Canadian Grocer*, 3 October 1919, 30; "Educate Your Trade – They'll Buy Fish," *Canadian Grocer*, 3 October 1919, 32; "Know More about Fish," *Canadian Grocer*, 3 October 1919, 50; "Sells Nearly Two Thousand Pounds of Halibut in This Store Every Week," *Canadian Grocer*, 20 January 1922, 23.

30 "Fish in Summer Diet," *Canadian Grocer*, 14 April 1905, 17.

31 "Good, Healthful, Nourishing Food," advertisement for Black Bros. & Co. Ltd, *Canadian Grocer*, 14 October 1904, 16.

32 "The Eating Qualities," advertisement for Black Bros. & Co. Ltd, *Canadian Grocer*, 21 October 1904, 114.

33 "Next to Your Habits," advertisement for Black Bros. & Co. Ltd, *Canadian Grocer*, 29 September 1905, 51.

34 "A Light, Nourishing Summer Food That Makes a Strong Appeal at This Time," advertisement for Connors Brothers Ltd, *Canadian Grocer*, 23 July 1915, 44. See also "Plan for Bigger Fish Sales This Summer," advertisement for Connors Brothers Ltd, *Canadian Grocer*, 18 May 1917, 56.

35 "The Fish Cake," advertisement for Black Bros. & Co. Ltd, *Canadian Grocer*, 28 October 1904, 14.

36 "This Is Not a Fad," advertisement for Black Bros. & Co. Ltd, *Canadian Grocer*, 26 November 1904, 16. Text in another company advertisement claimed, "There is no more wholesome and nourishing fish in the world than those selected specially for Brunswick Brand." See "A Delightful Change from the Continual Breakfast of Bacon and Eggs," advertisement for Connors Brothers Ltd, *Canadian Grocer*, 16 June 1915, 50.

37 "The Breakfast Table Problem," advertisement for Black Bros. & Co. Ltd, *Canadian Grocer*, 7 October 1904, 16.

38 "Pure Codfish," advertisement for Black Bros. & Co. Ltd, *Canadian Grocer*, 20 January 1905, 16.

39 "The Best Fish in the World Is 'Halifax' Prepared Codfish," advertisement for Black Bros. & Co. Ltd, *Canadian Grocer*, 24 November 1905,

57. Although Black Brothers was the most aggressive brand-name company in the first decade of the twentieth century, it was not alone in pushing the idea that fish was good brain food. Wholesaler J. Bowman & Co. argued that fish was "great brain food" in "Are You Handling the Great Economic Brain Food," advertisement for J. Bowman & Co. Wholesale Fish, Toronto, 28 January 1916, 48. Likewise, North Atlantic Fisheries Ltd noted that "the Consumer soon realizes its nutritive value" in their advertisement "Rush in Your Fish Orders for Lent," *Canadian Grocer*, 31 January 1913, 6.

40 Connors Brothers noted in one advertisement, "Fish is a very desirable food in all seasons – so easily digested, and healthful." See "Keeping Up Sales," advertisement for Connors Brothers Ltd, *Canadian Grocer*, 27 April 1915, 56.

41 "Discriminating People," advertisement for Brunswick Brand Sea Foods, Connors Brothers Ltd, *Canadian Grocer*, 11 January 1918, 52.

42 "Push Brunswick Brand Sea Foods for a Greater Lenten Profit," advertisement for Brunswick Brand Sea Foods, Connors Brothers Ltd, *Canadian Grocer*, 19 March 1920.

43 "Put Them in Your Window for Lent," advertisement for Connors Brothers Ltd, *Canadian Grocer*, 26 February 1926.

44 "Be Sure to Order a Large Stock of Brunswick Brands to Fill the Fall and Winter Demands," advertisement for Brunswick Brand Sea Foods, Connors Brothers Ltd, *Canadian Grocer*, 3 October 1919.

45 "Sea Food Profits," advertisement for Brunswick Brand Sea Foods, Connors Brothers Ltd, *Canadian Grocer*, 24 October 1919; "Read What the Food Board at Ottawa Has to Say about Brunswick Brand Fish Foods," advertisement, Connors Brothers Ltd, *Canadian Grocer*, 7 November 1919.

46 "Caught in Canadian Waters by Canadian Fishermen," advertisement for Connors Brothers Ltd, *Canadian Grocer*, 22 February 1915, 53.

47 "Sell the All-Canadian Sea Foods – Brunswick Brand," advertisement for Connors Brothers Ltd, *Canadian Grocer*, 30 July 1917, 54. See also "Packed for Canadians by Canadians," advertisement for Connors Brothers Ltd, *Canadian Grocer*, 11 January 1924, 43; "Profits – Which Have Honor in Their Own Country," advertisement for Connors Brothers Ltd, *Canadian Grocer*, 15 January 1924, 50.

48 A.H. Brittain, general manager, Maritime Fish Corporation Ltd, Montreal, to J.A. Choquette, Esq., Sherbrooke, Que., 21 November 1923, General

Information, Advertising Campaign by the Canadian Co-operative Fish Publicity Fund to Increase Demand for Fish, July 1923 to July 1924, LAC, RG 23, vol. 528, folder 711-25-16 (1).

49 "Are You Fit? If Not Here Is the Secret!" & "If You Work 'Inside' Eat More Fish," General Information, Advertising Campaign by the Canadian Co-operative Fish Publicity Fund to Increase Demand for Fish, July 1923 to July 1924, LAC, RG 23, vol. 528, folder 711-25-16 (1).

50 Although the 1920s have been referred to as the golden age of women's sports as it became more acceptable for women to engage in some athletic endeavours, food advertising related to athleticism still overwhelmingly focused on masculinity. See, for example, the Canadian Sports Hall of Fame at http://canadasports150.ca/en/women-in-canadian-sport/golden-age-of-women-in-sport/6.

51 See, for example, Kathryn Morse, *Nature of Gold: An Environmental History of the Klondike Gold Rush* (Seattle: University of Washington Press, 2010); and Jackson Lears, *Rebirth of a Nation: The Making of Modern America, 1877–1920* (New York: Harper Perennial, 2010).

52 "Eat Fish for Health," General Information, Advertising Campaign by the Canadian Co-operative Fish Publicity Fund to Increase Demand for Fish, July 1923 to July 1924, LAC, RG 23, vol. 528, folder 711-25-16 (1).

53 Mosby, *Food Will Win the War.*

54 Jack Doyle, *Altered Harvest: Agriculture, Genetics, and the Fate of the World's Food Supply* (New York: Penguin, 1985); Jeremy Rifkin, *Beyond Beef: The Rise and Fall of Cattle Culture* (New York: Penguin, 1993); Steven Stroll, *The Fruits of Natural Advantage: Making the Industry Countryside of California's Heartland* (Berkeley: University of California Press, 1998); Victoria Saker Woeste, *The Farmer's Benevolent Trust: Law and Agricultural Cooperation in Industrial America, 1865–1945* (Chapel Hill: University of North Carolina Press, 1998); Ian Tyrell, *True Gardens of the Gods: Californian-Austrian Environmental Reform, 1860–1930* (Berkeley: University of California Press, 1999); Marc Linder and Lawrence S. Zacharias, *Of Cabbage and King County: Agriculture and the Formation of Modern Brooklyn* (Iowa City: University of Iowa Press, 1999); David Vaught, *Cultivating California: Growers, Specialty Crops, and Labor, 1875–1920* (Baltimore: Johns Hopkins University Press, 2002); Roger Horowitz, *Putting Meat on the American Table: Taste, Technology, and Transformation* (Baltimore: Johns Hopkins University Press, 2005); Steve Striffler, *Chicken: The Dangerous Transformation of America's Favorite Food* (New Haven: Yale University Press, 2005); Douglas Cazaux Sackman, *Orange Empire: California and the Fruits of Eden* (Berkeley: University of

California Press, 2005); Ted Steinberg, *Down to Earth: Nature's Role in American History*, 2nd ed. (New York: Oxford University Press, 2009), 173–202.

55 "In Point of Tastiness," advertisement by Black Bros., Halifax, Nova Scotia, *Canadian Grocer*, 15 July 1904, 33.

56 "As Fresh and Clean as a Breeze from the Sea," advertisement for Connors Brothers Ltd, *Canadian Grocer*, 26 October 1917, 70; "Delicious as a Breeze from the Old Atlantic," advertisement for Brunswick Brand Sea Foods, Connors Brothers Ltd, *Canadian Grocer*, 25 October 1918.

57 "Decidedly Not!," advertisement for Connors Brothers Ltd, *Canadian Grocer*, 3 February 1922, 41.

58 Neil Forkey, *Canadians and the Natural Environment to the Twenty-First Century* (Toronto: University of Toronto Press, 2012).

59 Mildred Campbell, "Food and Health Value of Fish, 1930," Development of Fisheries: Advertising Fisheries through Radio Broadcasting, LAC, RG 23, vol. 528, folder 711-25-18 (1).

60 Department of Fisheries, "Food and Health Value of Fish," Development of Fisheries: Advertising Fisheries through Radio Broadcasting, LAC, RG 23, vol. 528, folder 711-25-18 (1).

61 Paisley was from Sackville, NB, and graduated from Mount Allison University. He was mainly involved in journalism and public relations and served as editor and editor-in-chief for newspapers before his employment with the Department of Marine and Fisheries. He gave numerous public addresses on behalf of the department, many of them announced in Ottawa newspapers. He was also very involved in the Mount Allison University alumni association (he even named his son Allison) and was given an honorary degree from that institution in 1954. He died in 1975. See his obituary in the *Ottawa Journal*, 23 January 1975, 44.

62 H.F.S. Paisley, Director, Fisheries Intelligence and Publicity Division, Department of Marine and Fisheries, "Canada's Fisheries Resources," for Broadcast under the Auspices of the Professional Institute of the Civil Service on March 25th, 1930, 8:45 to 9 pm, released for publication in Morning Newspapers on March 26th, Development of Fisheries: Advertising Fisheries through Radio Broadcasting, LAC, RG 23, vol. 528, folder 711-25-18 (1). Paisley came to the Department of Marine and Fisheries by way of the *Daily Recorder* (Sydney, NS), where he was editor. He was a public-relations expert, not a fisheries expert. He appeared frequently in Ottawa-area newspapers for talks he gave for the department to various social groups, often women's clubs, and for his active involvement with

the Mount Allison Alumni Association. See, for example, "War Publicity Problem Spotlighted in House," *National Post*, 10 August 1940, 1; "Report to House on Liaison Officers," *Ottawa Journal*, 6 August 1940, 15; "Social and Personal Activities," *Ottawa Citizen*, 18 December 1940, 5.

63 Radio Talk, Ottawa, February 2, 1932, Information, Radio Talks – Fisheries Campaign, Mrs. E. Spencer, 16 January 1932 to March 1938, LAC, RG 23, vol. 532, folder 711-25-23 (1).

64 "Sell Brunswick Brand Sea Foods for Profit and Reputation," advertisement for Connors Brothers Ltd, *Canadian Grocer*, 1 January 1, 1915, 56.

65 "This Is a Profit Maker," advertisement for Connors Brothers Ltd, *Canadian Grocer*, 12 May 1916, 58.

66 "The Fish You Sell Most Of," advertisement for Thistle Brand Fish, Arthur P. Tippet & Co., Agents, Montreal, *Canadian Grocer*, 16 June 16, 1915, 2.

67 "Fresh from the Sea! Packed in British Columbia," advertisement for Canadian Canners Ltd, Hamilton, and Canadian Canners (Western) Ltd, Vancouver, *Canadian Grocer*, 11 December 1936, 13.

68 "Thanks to This Modern Miracle Now Any Day Is Fish Day," advertisement for American Can Company, Vancouver, *Canadian Grocer*, 17 September 1937, 63.

69 "Selling Facts Worth Knowing about Canned Lobster," advertisement for American Can Company, Vancouver, in *Canadian Grocer*, 15 June 15, 1934, 1.

70 "Ways to Sell More Canned Salmon," advertisement for American Can Company, Vancouver, *Canadian Grocer*, 18 May 1934, 1.

71 "That's One Fish Story I'm Glad I Heard!," advertisement for American Can Company, *Canadian Grocer*, 15 September 1939, 56.

72 "Thanks to This Modern Miracle Now Any Day Is Fish Day," advertisement for American Can Company, Vancouver, *Canadian Grocer*, 17 September 1937, 63. Tea was also widely marketed as a brain-building ingredient. See, for example, "Tea Helped Discover the North Pole ... Drink More Tea for Vitality," *Canadian Home Journal*, May 1936, 81.

73 Ann Adam, "Fish – Any Time, Any Place," *Canadian Home Journal*, February 1932, 46, 50b.

74 Ann Adam, "We Suggest Fish for Autumn Menus," *Canadian Home Journal*, October 1936, 75.

75 "Let's Go Fishing: The Cooking Class Conducted by the Home Service Bureau," *Canadian Home Journal* 28, no. 9 (January 1932): 32–3, 36. Cod liver oil, of course, could be purchased as a stand-alone product, independent of the cod it came from. See for example, "Colds? ... Not One

All Winter," advertisement for Scott's Cod Liver Oil, *Canadian Home Journal*, March 1933, 51, specifically marketing the vitamin A and D content of cod liver oil as a means to prevent colds; and "Old Scottie," advertisement cartoon for Scott's Emulsion Cod Liver Oil, *Canadian Home Journal*, December 1937, 38, which marketed cod liver oil as a cure for rickets.

76   "Should the Children Eat It?," *Canadian Home Journal*, January 1932, 56.

77   "Mother Know Best … Does She?" *Canadian Home Journal*, March 1932, 88. See also Helen Gagen, "Vitamins Are Still in the News," *Canadian Home Journal*, June 1939, 50. In this piece, Gagen provides an overview of vitamins A, B1, C, D, E, and K but refers to the need of the "amateur star-gazer" to trust the professionals.

78   David M. Damkaer, "Mildred Helena Campbell (1907–2004): Early Copepodologist from British Columbia," *Journal of Crustacean Biology* 31, no. 4 (2011): 742–5.

79   "Food and Health Value of Fish," text of first radio addresses by M. Campbell, 1930, Development of Fisheries, Advertising Fisheries through Radio Broadcasting, LAC, RG 23, vol. 528, folder 711-25-18 (1).

80   H.F.S. Paisley, Director, Fisheries Intelligence and Publicity Division, Department of Marine and Fisheries, "Canada's Fisheries Resources," for Broadcast under the Auspices of the Professional Institute of the Civil Service on March 25th, 1930, 8:45 to 9 pm, released for publication in Morning Newspapers on March 26, Development of Fisheries, Advertising Fisheries through Radio Broadcasting, LAC, RG 23, vol. 528, folder 711-25-18 (1). The text was also published in full by the *Charlottetown Guardian*, 26 March 1930.

81   Wm. A. Found, D.Sc. Deputy Minister of Fisheries, "The Canadian Fisheries and the Work of the Department of Fisheries, for the Professional Institute of the Civil Service of Canada," Wednesday, 24 February 1932, CNRO, Ottawa, Development of Fisheries, Advertising Fisheries through Radio Broadcasting, LAC, RG 23, vol. 528, folder 711-25-18 (1).

82   "Opening Talk on Fish Cookery, over C.N.R.O., Broadcasted by M.E. Campbell, Tuesday, April 1st, 1930," Development of Fisheries, Advertising Fisheries through Radio Broadcasting, LAC, RG 23, vol. 528, folder 711-25-18 (1).

83   "Talk on Fish, Given for the Home Service Department of the Ottawa Electric & Gas Company, Ltd., Sparks Street, Ottawa, by M.E. Campbell," Development of Fisheries, Advertising Fisheries through Radio Broadcasting, LAC, RG 23, vol. 528, folder 711-25-18 (1).

84   Wm. A. Found, D.Sc., Deputy Minister of Fisheries, "The Canadian Fish-

eries and the Work of the Department of Fisheries, for the Professional Institute of the Civil Service of Canada," Wednesday, 24 February 1932, CNRO, Ottawa, Development of Fisheries, Advertising Fisheries through Radio Broadcasting, LAC, RG 23, vol. 528, folder 711-25-18 (1).

85 "Script for Radio Talk, Ottawa, Feb 4, 1932," Information, Radio Broadcast, 25 June 1930 to 7 February 1934, LAC, RG 23, vol. 528, folder 711-25-18 (2).

86 Many of these broadcasts can be found in General Information, Radio Broadcasts, 7 November 1934 to 25 October 1937, LAC, RG 23, vol. 528, folder 711-25-18 (4).

87 "Every Day Is Friday," B.E. Bailey, Pacific Fisheries Experimental Station, Prince Rupert, BC, n.d., General Information, Radio Broadcasts, 7 November 1934 to 25 October 1937, LAC, RG 23, vol. 528, folder 711-25-18 (4).

88 Dr W.A. Clemens, Director, Biological Station, Nanaimo, BC, "The Sea as Nature's Storehouse of Food Materials," for Broadcast, Friday, 21 December 1934, 9:15–9:30 PM, C.R.C.V., Vancouver, for Newspaper Release, 22 December 1934, General Information, Radio Broadcasts, 7 November 1934 to 25 October 1937, LAC, RG 23, vol. 528, folder 711-25-18 (4).

89 Clemens presented a confusing picture of the marine food chain that suggested that the fish humans ate were plant grazers.

90 Dr W.A. Clemens, Director, Biological Station, Nanaimo, BC, "The Sea as Nature's Storehouse of Food Materials," for broadcast, Friday, 21 December 1934, 9:15–9:30 PM, C.R.C.V. Vancouver, for Newspaper Release, 22 December 1934, General Information, Radio Broadcasts, 7 November 1934 to 25 October 1937, LAC, RG 23, vol. 528, folder 711-25-18 (4).

91 Michaud was elected as a Liberal MP for Restigouche-Madawaska in 1933 and served as minister of fisheries from 1935 to 1945. He also served as minister of transportation from 1942 to 1945, as acting minister of justice and attorney general of Canada in 1941, and acting minister of public works in 1942.

92 Transcript of Broadcast by the Minister from the Nova Scotia Hotel, Halifax, 20 October 1937, 8:45–9:00 PM, General Information, Radio Broadcasts, 7 November 1934 to 25 October 1937, LAC, RG 23, vol. 528, folder 711-25-18 (4).

93 "A Crusade That Will Multiply Every Canadian Dealer's Profit," advertisement by Department of Fisheries, *Canadian Grocer*, 6 September 1936, 29; "Such Flavour! Serve Fish Often," advertisement by Department of Fisheries, *Canadian Home Journal*, October 1936, 74; "Add Your Windows to the Parade," advertisement by Department of Fisheries,

Canadian Grocer, 13 November 1936, 16; "Display, Recommend, Sell
Fish and Shellfish," advertisement by Department of Fisheries, *Canadian
Grocer*, 3 January 1937, 25; "Are You Profiting by the National Trend to
Fish?," advertisement by Department of Fisheries, *Canadian Grocer*, 3
March 1937, 16; "Another Great Fish Campaign Is on the Way," adver-
tisement by Department of Fisheries, *Canadian Grocer*, 15 February
1938, 31; "Here Is Your Market for Fish," advertisement by Department
of Fisheries, *Canadian Grocer*, 1 October 1938, 4; "Make Every Day a
Fish Day," advertisement by Department of Fisheries, *Canadian Grocer*, 1
November 1938, 19; "The Third Great Fish Campaign Is Now Appearing All over Canada," advertisement by Department of Fisheries, Ottawa,
Canadian Grocer, 1 February 1939, 23; "People Must Be Eatin' an Awful
Lot of Fish ...," advertisement for Department of Fisheries, Ottawa, *Canadian Grocer*, 15 September 1939, 17; "What Makes It Ring So Often, Mis-
ter? Mostly Fish, Sonny!," advertisement by Department of Fisheries,
Ottawa, *Canadian Grocer*, 1 February 1940, 12; all these ads were part of
the Any Day a Fish Day marketing by the department.

94 Atlantic Advertising Agency to H.M.F. Paisley, Director of Intelligence
and Publicity, Department of Fisheries, Ottawa, 4 February 1937, Expansion of Fishing Trade, 1 February 1937 to 13 February 1937, LAC, RG 23,
vol. 533, folder 711-25-24 (13).

95 "Serve Fish for Family Health," Any Day a Fish Day, Department of
Fisheries, proof located in Expansion of Fishing Trade, Advertising,
19 October 1937 to 9 November 1937, LAC, RG 23, vol. 534, folder 711-
25-24 (19).

96 "Try This Delicious recipe: Fish Is Nourishing, Flavoursome, Economi-
cal," advertisement of Department of Fisheries, *Evening Empire* (Prince
Rupert, BC), 21 October 1937; "The Vitamin Fish," advertisement for
Canadian Fish and Cold Storage Co. Ltd, *Evening Empire* (Prince
Rupert, BC), 21 October 1937.

97 "Radiantly Healthy ... Fish Helps Keep Them So," advertisement for
Department of Fisheries, *Sutana Herald* (Saskatoon, SK), 21 December
1938.

98 "Pour les écoliers le Poisson est un Nourriture précieuse," advertisement
for Ministère De Pêcheries, Ottawa, Le Poisson au menu du jour, *La
Revue Populaire* 32, no. 10 (October 1939): 53.

99 "Schools Out!," advertisement proof sent from W. George Akins, Walsh
Advertising, Toronto, to Department of Fisheries, Advertising Fisheries,
Generally, 22 June 1939 to 21 August 1939, LAC, RG 23, vol. 536, folder
711-25-24 (34).

100 "Strong Bodies, Sound Teeth, Good Growth, Promoted by Fish," *Hamilton Spectator*, 21 October 1936, 13; "Serve Fish for Family Health," Any Day a Fish Day, Department of Fisheries, Expansion of Fishing Trade, advertising, 19 October 1937 to 9 November 1937, LAC, RG 23, vol. 534, folder 711-25-24 (19). Similar advertisements appeared in the French-language press; see "Pour force et santé servez du poisson," advertisement, Du poisson n'importe quel jour, Advertising the Fisheries, General, 25 January 1938 to 18 February 1938, LAC, RG 23, vol. 535, folder 711-25-38 (23).

101 Ad proofs sent from E.W. Reynolds, December 1937, to H.M.F. Paisley, Director of Publicity, Department of Fisheries, Expansion of Fishing Trade, Advertising, 23 November 1937 to 4 January 1938, LAC, RG 23, vol. 534, folder 711-25-24 (21).

102 "A Message to Men ... Eat Fish for Health," advertisement for Any Fish a Fish Day, Department of Fisheries, *Montreal Herald*, 9 March 1938; "A Message to Men ... Eat Fish for Health," advertisement for Any Fish a Fish Day, Department of Fisheries, *Star Weekly* (Toronto), 12 March 1938, 12.

103 E.W. Reynolds & Co. Ltd, Copy of text intended for Jewish papers, 2 September 1936, Expansion of Fish Trade, Generally, 8 August 1936 to 10 September 1936, LAC, RG 23, vol. 532, folder 711-25-24 (3).

104 See, for example, "Si vous voulez un aliment sain, savoureux, nutritif, économique, mangez du poisson!," Du poisson n'importe quel jour advertisement for the Ministère des Pêcheries, *L'Illustration nouvelle*, 27 October 1937, 7. "Digne d'un roi!," Du poisson n'importe quel jour, *Le Club Automobile de Quebec*, n.d.; "De nouvelles recettes de poisson font le régal de la famille," Le Poisson au menu du jour advertisement, *La Famille* (Montreal), November 1938; "Pour les écoliers le poisson est un nourriture précieuse," Le Poisson au menu du jour advertisement, *La Revue Populaire* 32, no. 10 (October 1939): 53.

105 Katherine Caldwell, Home Service Bureau, "Cooking Class Supplement," *Canadian Home Journal*, October 1937.

106 See, for example, *Star Weekly* (Toronto), 12 March 1938. Such editorials appeared elsewhere as well; see "Les repas de poisson sont faciles a préparer," *L'Action paroissiale*, February 1940, 19.

107 The original drafts and final proof of this message can be found in Expansion of Fishing Trade, Advertising, 23 November 1937 to 4 January 1938, LAC, RG 23, vol. 534, folder 711-25-24 (21).

108 "Thanks to This Modern Miracle, Now Any Day Is Fish Day," advertisement for American Can Company, Vancouver, *Canadian Grocer*, 17 September 1937, 63.

109 "That's One Fish Story I'm Glad I Heard!," advertisement for American Can Company, *Canadian Grocer*, 15 September 1939, 56.

110 "The Greatest Sardine Value," advertisement for Connors Brothers Ltd, *Canadian Grocer*, 25 March 1932, cover page.

111 "Headed by Brunswick Brand," advertisement for Connors Brothers Ltd, *Canadian Grocer*, 29 July 1932, cover page.

112 "It Pays to Display," advertisement for Connors Bros. Famous Sea Foods, Connors Brothers Ltd, *Canadian Grocer*, 5 April 1935, 3; "For Quick Sales and Profits," advertisement for Connors Bros. Famous Sea Foods, Connors Brothers Ltd, *Canadian Grocer*, 14 June 1935, 5; "Canadian Fish … For Canadian Tables," advertisement for Connors Bros. Famous Sea Foods, Connors Brothers Ltd, *Canadian Grocer*, 5 August 1935, 5; "Carry the Complete Line of Connors Bros. Famous Sea Foods," advertisement for Connors Bros. Famous Sea Foods, Connors Brothers Ltd, *Canadian Grocer*, 18 October 1935, 81; "12 Fastest Selling Sea Foods," advertisement for Connors Bros. Famous Sea Foods, Connors Brothers Ltd, *Canadian Grocer*, 2 April 1936, 7; "No Shelf Warmers!," advertisement for Connors Bros. Famous Sea Foods, Connors Brothers Ltd, *Canadian Grocer*, 12 June 1936, 5; "Accent on Profit," advertisement for Connors Bros. Famous Sea Foods, Connors Brothers Ltd, *Canadian Grocer*, 11 June 1937, 3.

113 J.H. Todd & Sons was one such company that tried to use the government's general advertising to boost sales of their brand. The company, along with the American Can Company, put out extensive advertising utilizing print and radio to market their canned salmon. See, for example, "3 Big Reasons Why You Should Stick and Push the Sale of Todd's Horse Shoe Brand Fancy Red Sockeye Salmon," advertisement for J.H. Todd & Sons Ltd, Victoria, BC, *Canadian Grocer*, 3 February 1937, 2; "A High Quality Established Brand Backed by Intensive Consumer Advertising," advertisement for J.H. Todd & Sons Ltd, Victoria, *Canadian Grocer*, 3 March 1937, 38; "Another Big Advertising Campaign for Todd's Canned Salmon," advertisement for J.H. Todd & Sons Ltd, Victoria, *Canadian Grocer*, 11 June 1937, 35; "Thanks to This Modern Miracle [...]," advertisement for American Can Company, Vancouver, *Canadian Grocer*, 17 September 1937, 63. See also "Pacific Halibut 'The Vitamin Fish,'" advertisement for Canadian Fish & Cold Storage Co. Ltd, *Canadian Fisherman*, 26, no. 6 (June 1939): 5.

114 R.E. Wodehouse, M.D., Deputy Minister, Department of Pensions and National Health, to J.J. Cowie, Acting Deputy Minister, Department of

Fisheries, 23 January 1940, Advertising Fisheries, Generally, 20 December 1939 to 21 February 1940, LAC, RG 23, vol. 537, folder 711-25-24 (38).

115 J.J. Cowie, Acting Deputy Minister, Department of Fisheries, to R.E. Wodehouse, M.D., Deputy Minister, Department of Pensions and National Health, to 27 January 1940, Advertising Fisheries, Generally, 20 December 1939 to 21 February 1940, LAC, RG 23, vol. 537, folder 711-25-24 (38).

116 J.M. Bowman, Walsh Advertising Co. Ltd, Toronto, to H.F.S. Paisley, Director of Publicity, Department of Fisheries, Ottawa, 10 July 1939, Advertising Fisheries, Generally, 22 June 1939 to 21 August 1939, LAC, RG 23, vol. 536, folder 711-25-24 (34).

117 Report of E.W. Reynold & Co. Ltd to Honourable J.E. Michaud, Department of Fisheries, Ottawa, 10 February 1939, 3, Advertising Fisheries, Generally, 7 February 1939 to 28 February 1939, LAC, RG 23, vol. 536, folder 711-25-24 (31).

118 R.E. Wodehouse, M.D., Deputy Minister, Department of Pensions and Public Health, to J.J. Cowie, Acting Deputy Minister, Department of Fisheries, 16 February 1940, Advertising Fisheries, Generally, 5 February 1940 to 18 April 1940, LAC, RG 23, vol. 537, folder 711-25-24 (39).

CHAPTER THREE

1 There is no evidence in the archival record of any effort by the federal government or the industry to market seafood to Canada's non-white or Indigenous populations; neither does any advertising or marketing make a distinction between Anglo and French Canadian customers. There are occasional references to religion, as marketers often crafted advertisements for Catholic and Jewish populations separate from their main messaging.

2 Sherrie A. Inness, ed., *Cooking Lessons: The Politics of Gender and Food* (New York: Rowman & Littlefield, 2001), xii.

3 Patricia M. Gantt, "Taking the Cake: Power Politics in Southern Life and Fiction," in Inness, *Cooking Lessons*, 63.

4 Jessamyn Neuhaus, "Is Meatloaf for Men? Gender and Meatloaf Recipes, 1920–1960," in Inness, *Cooking Lessons*, 89–90.

5 Examples of the extensive historiography on advertising include Laura Baker, "Public Sites versus Public Sights: The Progressive Response to Outdoor Advertising and the Commercialization of Public Space," *American Quarterly* 59, no. 4 (December 2007): 1187–1213; Susan Porter

Benson, *Counter Cultures: Saleswomen, Managers, and Customers in American Department Stores, 1890–1940* (Chicago: University of Illinois Press, 1986); Richard Wightman Fox and T.J. Jackson Lears, eds., *The Culture of Consumption: Critical Essays in American History, 1880–1980* (New York: Pantheon Books, 1983); Russell Johnston, *Selling Themselves: The Emergence of Canadian Advertising* (Toronto: University of Toronto Press, 2011); T.J. Jackson Lears, *No Place of Grace: Antimodernism and the Transformation of American Culture, 1880–1920* (New York: Pantheon Books, 1989); T.J. Jackson Lears, *Fables of Abundance: A Cultural History of Advertising in America* (New York: Basic Books, 1994); William R. Leach, "Transformations in a Culture of Consumption: Women and Department Stores, 1890–1925," *Journal of American History* 71, no. 2 (September 1984): 319–42; Lori Anne Loeb, *Consuming Angels: Advertising and Victorian Women* (New York: Oxford University Press, 1994); Roland Marchand, *Advertising the American Dream: Making Way for Modernity, 1920–1940* (Berkeley: University of California Press, 1985); James D. Marchand, *Advertising and the Transformation of American Society, 1865–1920* (Westport, CT: Greenwood, 1990); Richard Ohmann, *Selling Culture: Magazines, Markets and Class at the Turn of the Century* (London: Verso, 1996); Michael Schudson, *Advertising, the Uneasy Persuasion: Its Dubious Impact on American Society* (New York: Basic Books, 1984); Susan Strasser, *Satisfaction Guaranteed: The Making of the American Mass Market* (New York: Pantheon, 1989); Richard S. Tedlow, *New and Improved: The Story of Mass Marketing in America* (New York: Basic Books, 1990); Priscilla J. Brewer, "'We Have Got a Very Good Cooking Stove': Advertising, Design, and Consumer Response to the Cookstove, 1815–1880," *Winterthur Portfolio* 25, no. 1 (1990): 35–54.

6 Katherine J. Parkin, *Food Is Love: Advertising and Gender Roles in Modern America* (Philadelphia: University of Pennsylvania Press, 2006), 8.

7 Parkin, *Food Is Love*, 8–9.

8 Valerie J. Korinek, *Roughing It in the Suburbs: Reading* Chatelaine *Magazine in the Fifties and Sixties* (Toronto: University of Toronto Press, 2000), 15.

9 Ibid., 11.

10 Ibid., 8.

11 Ibid., 14.

12 Ibid., 122.

13 Laura Shapiro uses the term "domestic theology" to describe the moral value that women associated with their domestic work following industrialization. See *Perfection Salad: Women and Cooking at the Turn of the*

Century (New York: Farrar, Straus and Giroux, 1986), 21. Shapiro argues that there was a strong sense of moralism in women's domestic work during the late industrial period into the twentieth century. She often states that women saw this as "divine" in nature but does not suggest that this was universal, instead focusing on how popular media presented the argument that women ought to consider their work divine. This propaganda appeared in cookbooks and food advertisement and is a useful analytical tool for this study.

14 Andrew Holman, *A Sense of Their Duty: Middle-Class Formation in Victorian Ontario Towns* (Montreal and Kingston: McGill-Queen's University Press, 2000), 7.

15 Ibid., ix.

16 Ibid.

17 Paul Axelrod, *Making a Middle Class: Student Life in English Canada during the Thirties* (Montreal and Kingston: McGill-Queen's University Press, 1990), 168.

18 Ibid., 168.

19 Ibid., 169.

20 Ibid., 170.

21 Ibid., 7.

22 Franca Iacovetta, "The Gatekeepers: Middle-Class Campaigns of Citizenship in Early Cold War Canada," in *The Making of the Middle Class: Toward a Transnational History*, ed. A. Ricardo Lopez and Barbara Weinstein (Durham, NC: Duke University Press, 2012), 87. See also Franca Iacovetta, "Immigrant Gifts, Canadian Treasures, and Spectacles of Pluralism: The International Institute of Toronto in North American Context, 1950s–1970s," *Journal of American Ethnic History* 31, no. 1 (2011): 34–73.

23 Iacovetta, "Gatekeepers," 92.

24 Shapiro, *Perfection Salad*, 5–7, 72, 90.

25 Ibid., 203–6, 214.

26 Ibid., 9.

27 Mary Needler Arai, a pioneer woman in fisheries science in Canada herself, reviews some of the exceptional women who provided leadership in the Department of Fisheries in her "Some Contributions of Women to the Early Study of the Marine Biology of Canadian Waters," in *A Century of Maritime Science: The St. Andrews Biological Station*, ed. Jennifer Hubbard, David Wildish, and Robert Stephenson (Toronto: University of Toronto Press, 2016): 50–77.

28 Matthew McKenzie, "Iconic Fishermen and the Fates of New England Fisheries Regulations, 1883–1912," *Environmental History*, no. 17 (January 2012): 15.

29 While women did make important contributions to the labour of fishing, especially in the cannery industry, the popular imagery of fisheries, including that utilized in advertising, always depicted the fishing labourer as male.

30 For a gender history of consumerism in twentieth-century America, see Benson, *Counter Cultures*; Brewer, "'We Have Got a Very Good Cooking Stove,'" 35–54; Ellen Gruber Garvey, "Reframing the Bicycle: Advertising-Supported Magazines and Scorching Women," *American Quarterly* 74, no. 1 (1995): 66–101; William R. Leach, "Transformations in a Culture of Consumption: Women and Department Stores, 1890–1925," *Journal of American History* 71, no. 2 (1984): 319–42; Barbara Penner, "'A Vision of Love and Luxury': The Commercialization of Nineteenth-Century American Weddings," *Winterthur Portfolio* 39, no. 1 (2004): 1–20; Linda L. Tyler, "'Commerce and Poetry Hand in Hand': Music in American Department Stories, 1880–1930," *Journal of the American Musicological Society* 45, no. 1 (1992): 75–120.

31 "Canned Sea Foods of Established Quality Mean Year Round Sales in Your Fish Department," advertisement for Connors Brothers Ltd, *Canadian Grocer*, 16 April 1915, 45.

32 "Thistle Brand," advertisement for Thistle Brand, Arthur P. Tippet & Company, Agents, Montreal, *Canadian Grocer*, 20 August 1915, 2.

33 "The Fish You Sell Most of," advertisement for Thistle Brand Fish, *Canadian Grocer*, 16 June 16, 1915, 2.

34 "Your Surest Guarantee," advertisement for Connors Brothers Ltd, *Canadian Grocer*, 15 January 1915, 56.

35 "King Oscar Sardines All Year Round," advertisement for John W. Bickle & Greening, Hamilton, Ont., *Canadian Grocer*, 13 January 1922, 48.

36 "A Delightful Change from the Continual Breakfast of Bacon and Eggs," advertisement for Connors Brothers Ltd, *Canadian Grocer*, 16 June 1915, 50.

37 "Such Flavour! Serve Fish Often," advertisement by Department of Fisheries, *Canadian Home Journal*, October 1936, 74.

38 "Number 8: "Ladies! Attention Please!" included in Atlantic Advertising Agency, Sackville, NB, to H.F.S. Paisley, Department of Fisheries, Ottawa, 20 October 1936, Expansion of Fishing Trade, 21 October 1936 to 26 October 1936, LAC, RG 23, vol. 533, folder 711-25-24 (7). See also J.L.

Hart, Pacific Biological Station, "The British Columbia Pilchard," 4 January 1935, General Information, Radio Broadcasts, 7 November 1934 to 25 October 1937, LAC, RG 23, vol. 528, folder 711-25-18 (4).

39 "The High Price of Meat Is Turning," advertisement for Connors Brothers Ltd, *Canadian Grocer*, 11 August 1916, 42.

40 "Help to Cut the Cost of Living," advertisement for Connors Brothers Ltd, *Canadian Grocer*, 14 September 1917, 78.

41 "Economical and Good," advertisement for Connors Brothers Ltd, *Canadian Grocer*, 23 November 1917, 52.

42 "Be Sure to Order a Large Stock of Brunswick Brands to Fill the Fall and Winter Demands," advertisement for Brunswick Brand Sea Foods, Connors Brothers Ltd, *Canadian Grocer*, 3 October 1919, emphasis in original.

43 "The Brand That's Easier to Sell," advertisement for Connors Brothers Ltd, *Canadian Grocer*, 8 February 1924, 51.

44 See also "Now – This Summer Improve Your Sardine Turnover with Banquet Brand Sardines," advertisement for Lewis Connors & Sons, West Saint John, New Brunswick, *Canadian Grocer*, 25 July 1924, 58. In "On the Family Menu," advertisement for Connors Brothers Ltd, *Canadian Grocer*, 28 February 1930, the firm celebrated the fact that "her economy induces the housewife to serve them regularly and often."

45 "Packed for Canadians by Canadians," advertisement for Connors Brothers Ltd, *Canadian Grocer*, 11 January 1924, 43.

46 Text from these radio commercials was included in W. Akins, E.W. Reynolds & Co. Ltd, to H.F.S. Paisley, Director of Publicity, Department of Fisheries, Ottawa, 19 January 1937, Expansion of Fishing Trade, 12 January 1937 to 30 January 1937, LAC, RG 23, vol. 533, folder 711-25-24 (12).

47 Arthur E. Chant, Esq., Regina, Sask., to Department of Fisheries, Ottawa, 8 February 1939, LAC, RG 23, vol. 511, folder 711-1-51 (5).

48 Mrs Dana F. Lucy, Calgary, to Hon. J.E. Michaud, 23 July 1940, LAC, RG 23, vol. 511, folder 711-1-51 (5).

49 W.S. Froop, Digby, Nova Scotia, to Hon. J.E. Michaud, Minister of Fisheries, Ottawa, 3 August 1940; W.L. Seely, Woodstock, N.B., to Department of Fisheries, Ottawa, 19 August 1940; F.G. Lamb, Esq., South Hamilton, Ont., to Department of Fisheries, Ottawa, 12 August 1940; P. Belleau, Quebec City, to the Hon. Minister of National Fisheries, Ottawa, 30 September 1940; Mrs A.N Lynn, Mayfair, Sask., to Mr J.E. Michaud, Minister of Fisheries, Ottawa, 16 March 1941; Mrs W.G. Beach, Clinton, BC, to Department of Fisheries, 10 June 1941; all in LAC, RG 23, vol. 511, folder 711-1-51 (5).

50 "The Secret History of the Week," *Halifax Herald*, 11 December 1937.

51 "The Best Way to Sell Fish," advertisement for Connors Brothers Ltd, *Canadian Grocer*, 28 September 1917, 54.

52 "Canada's Fish Education," advertisement for Connors Brothers Ltd, *Canadian Grocer*, 21 September 1917, 54.

53 "A Light, Nourishing Summer Food That Makes a Strong Appeal at This Time," advertisement for Connors Brothers Ltd, *Canadian Grocer*, 23 July 1915, 44.

54 "The Four Big Leaders Brunswick Brand Sea Foods," advertisement for Connors Brothers Ltd, *Canadian Grocer*, 5 May 1922, 54–55. See also "For Dainty, Easily Prepared Lunches, for Picnics, Etc., Suggest These Popular Brunswick Brand Sea Foods," advertisement for Connors Brothers Ltd, *Canadian Grocer*, 2 June 1922, 47; "Brunswick Brand Sea Foods Ready-to-Serve on Opening the Tin," advertisement for Connors Brothers Ltd, *Canadian Grocer*, 30 June 1929, 45. Connors Brothers also tried to push sardines into the breakfast meal. See, for example, "A Delightful Change from the Continual Breakfast of Bacon and Eggs," advertisement for Connors Brothers Ltd, *Canadian Grocer*, 16 June 1915, 50. In "Helping Your Customers – Helping Yourself," advertisement for Connors Brothers Ltd, *Canadian Grocer*, 18 July 1930, 28, the company suggested, "How much easier it is to slip a few tins of sardines into the picnic hamper, boys' knapsack or serve them for lunch, than to bother with countless sandwiches and complicated salads."

55 "A Meal in a Jiffy," advertisement for Connors Brothers Ltd, *Canadian Grocer*, 15 February 1924, 41. See also "Picnicers, Campers, and Cottagers Buy Connors' Sardines by the Half Case," advertisement for Connors Brothers Ltd, *Canadian Grocer*, 13 June 1924, 43.

56 "These Are Sardine Days," advertisement for Connors Brothers Ltd, *Canadian Grocer*, 18 June 1926, back of cover. See also "Do Your Profits Fall Off in the Summer Months?," advertisement for Connors Brothers Ltd, *Canadian Grocer*, 9 July 1926, 12; "For Summer Days," advertisement for Connors Brothers Ltd, *Canadian Grocer*, 20 July 1928, 11; "When Your Customers Go to the Country," advertisement for Connors Brothers Ltd, *Canadian Grocer*, 20 June 1930, 8.

57 Text included in Florence Robertson to F. Nelson, McConnell & Ferguson Ltd, 3 April 1931, Advertising Campaign, Requested by Canadian Fisheries Association, Generally, February 1931 to 16 July 1931, LAC, RG 23, vol. 531, folder 711-25-19 (3).

58 H.F.S. Paisley, Director of Fisheries Intelligence and Publicity, Department of Fisheries, Ottawa, to E.W. Reynolds & Co., Ltd, Toronto, 24

October 1936, Expansion of Fishing Trade, 21 October 1936 to 26 October 1936, LAC, RG 23, vol. 533, folder 711-25-24 (7).

59 "Advertising Our Fish," *Saint John Telegraph Journal*, 13 July 1928.

60 B.K. Snow, Chairman, Publicity Committee, Canadian Fisheries Association, Halifax, to F.T James, F.T. James Company, Toronto, 31 March 1926, General Information, Advertising Campaign by the Canadian Co-operative Fish Publicity Fund, LAC, RG 23, vol. 528, folder 711-25-16 (3).

61 Wm. A. Found, Deputy Minister of Fisheries, Ottawa, to J.J. Kinley, MP, Lunenburg, 2 December 1936, Expansion of Fishing Trade, 23 November 1936 to 16 December 1936, LAC, RG 23, vol. 533, folder 711-25-24 (10).

62 "Eat More Codfish," included in Letters from Canadian Fisheries Association to W.A. Found, Supt. of Fisheries and Dept. of Naval Services, March 1919, 31 January 1919 to 22 November 1919, LAC, RG 23, vol, 509, folder 711-1-33.

63 Wm. A. Found, Deputy Minister to Wm. Foran, Esq., Secretary, Civil Service Commission, 22 December 1931, "Advertising Fish – Mrs. E. Spencer – Lecturer – Demonstrator," LAC, RG 23, vol. 67, folder 711-1-2-1 (1).

64 William A. Found, D.Sc., Deputy Minister of Fisheries, "The Canadian Fisheries and the Work of the Department of Fisheries," for broadcast under the auspices of the Professional Institute of the Civil Service of Canada, Wednesday, 24 February 1932, CNRO, Ottawa, Development of Fisheries, Advertising Fisheries through Radio Broadcasting, LAC, RG 23, vol. 528, folder 711-25-18 (1).

65 Wm. A. Found, Deputy Minister of Fisheries, to the Manager, Canadian National Railway Broadcasting Station, Ottawa, 8 October 1929, Development of Fisheries. Advertising Fisheries through Radio Broadcasting, LAC, RG 23, vol. 528, folder 711-25-18 (1).

66 J.G. McMurtrie, Broadcasting Manager, Canadian National Railway, Radio Department, Ottawa, to Wm. A. Found, Deputy Minister of Fisheries, 12 October 1929, Development of Fisheries, Advertising Fisheries through Radio Broadcasting, LAC, RG 23, vol. 528, folder 711-25-18 (1).

67 M.I. Plaxton, Radio Specialists of Canada, Ltd, Toronto, to the Deputy Minister of Marine and Fisheries, 17 January 1930, Development of Fisheries. Advertising Fisheries through Radio Broadcasting, LAC, RG 23, vol. 528, folder 711-25-18 (1).

68 Open letter from C.A. Burkhardt, Past President of the Association of Pacific Fisheries and Member of the Advertising Committee, 7 February 1929, Development of Fisheries, Advertising Fisheries through Radio Broadcasting, LAC, RG 23, vol. 528, folder 711-25-18 (1).

69 M.E. Campbell, "Opening talk on Fish Cookery, over C.N.R.O.," 1 April
    1930, Development of Fisheries, Advertising Fisheries through Radio
    Broadcasting, LAC, RG 23, vol. 528, folder 711-25-18 (1).

70 M.E. Campbell, "Talk on Fish Cookery, Broadcast from C.N.R.O.," 8
    April 1930, Development of Fisheries, Advertising Fisheries through
    Radio Broadcasting, LAC, RG 23, vol. 528, folder 711-25-18 (1).

71 M.E. Campbell, "Talk in Fish Cookery broadcast over C.N.R.O.," 15
    April 1930, Development of Fisheries, Advertising Fisheries through
    Radio Broadcasting, LAC, RG 23, vol. 528, folder 711-25-18 (1).

72 M.E. Campbell, "Talk on Fish, Given for the Home Service Department
    of the Ottawa Electric & Gas Company, Ltd., Sparks Street, Ottawa,"
    Information, Radio Broadcast, 25 June 1930 to 7 February 1934, LAC, RG
    23, vol. 528, folder 711-25-18 (2).

73 "Sales of Fish Run over 1,500 Pounds a Week," *Canadian Grocer*, 15 Janu-
    ary 1924, 19.

74 Wm. A. Found, D.Sc., Deputy Minister of Fisheries, "The Canadian Fish-
    eries and the Work of the Department of Fisheries," for Broadcast under
    the auspices of the Professional Institute of the Civil Service of Canada,
    Wednesday, 24 February 1932, C.N.R.O., Ottawa, Information, Radio
    Broadcast, 25 June 1930 to 7 February 1934, LAC, RG 23, vol. 528, folder
    711-25-18 (2).

75 Evelene Spencer to William Found, Deputy Minister of Fisheries,
    Ottawa, 21 July 1931, Advertising Fish – Mrs. E. Spencer – Lecturer –
    Demonstrator, 27 April 1931 to 17 February 1932, LAC, RG 23, vol. 67,
    folder 711-1-2 (1).

76 See, for example, "Fish Is Boon to Busy Housewife," *Gazette* (Montreal),
    28 September 1933.

77 Civil Service Commission, Notification to Department (for Temporary
    Employment Only), Mrs Eveline Spencer, 16 February 1932; Wm. A.
    Found, Deputy Minister, to Mrs Evelene Spencer, Portland, Oregon, 16
    December 1931, Advertising Fish – Mrs E. Spencer – Lecturer – Demon-
    strator, LAC, RG 23, vol. 67, folder 711-1-2 (1). See also Wm. A. Found,
    Deputy Minister, to R.M. Winslow, Esq., Secretary, Canned Fish Section,
    Canadian Manufacturers' Association, Vancouver, BC, 2 February 1932,
    Advertising Fish – Mrs E. Spencer – Lecturer – Demonstrator, LAC, RG
    23, vol. 67, folder 711-1-2 (1).

78 After working for the US Bureau of Fisheries, Spencer had some success
    working for the National Fish Company of Halifax to promote the use
    of fresh fish fillet caught by their trawler fleet. See "Expert Says Judi-
    cious Use of Salt Is Secret of Cooking Fish," *Boston Herald*, 22 May 1919;

"Women Express Wish for Further Lessons Proper Way to Cook Fish," *Montreal Herald*, 28 November 1925; "Cook Fish to Be Delectable, Men Will Change Their Diet," *Toronto Star*, 29 December 1925; "Toronto Is in Great Luck over Chances for Fish," *Toronto Evening Telegram*, 27 February 1926; "Mrs. Spencer Asked to Continue Her Lecture," *Hamilton Herald*, 25 March 1926.

79 "Cook Fish to Be Delectable, Men Will Change Their Diet," *Toronto Star*, 29 December 1925.

80 "Fish Is Boon to Busy Housewife," *Gazette* (Montreal), 28 September 1933.

81 Spencer's reports are detailed in problems in her travels rather than the content of her demonstrations. She wrote in a friendly, airy way, often long-winded. Advertising Fish – Mrs E. Spencer – Lecturer – Demonstrator, LAC, RG 23, vol. 67, folders 711-1-2 (1) through 711-1-2 (4).

82 Radio Talk, Ottawa, 26 January 1932, Information, Radio Talks – Fisheries Campaign, Mrs. E. Spencer, 26 January 1932 to March 1938, LAC, RG 23, vol. 532, folder 711-25-23 (1).

83 "Summary of Report on the Marketing of Canadian Fish and Fish Products, to the Honourable E.N. Rhodes, Minister of Fisheries," by Cockfield, Brown & Company Ltd, Montreal, Toronto, Winnipeg, Vancouver, 1931, Expansion of Fish Trade, Survey of Fishing Industry, 19 January 1932, LAC, RG 23, vol. 531, folder 711-25-22 (3).

84 "Radio Talk, Ottawa, February 16, 1932," Information, Radio Broadcast, 25 June 1930 to 7 February 1934, LAC, RG 23, vol. 528, folder 711-25-18 (2). Emphasis in original.

85 "Mrs. Evelene Spencer," *Canadian Fisherman* 22, no. 2 (January 1935): 13.

86 Evelene Spencer, "Mrs. Evelene Spencer Tells Us 'Some of the Things I Have Come Across in the Fishing Industry,'" *Canadian Fisherman* 21, no. 1 (January 1934): 1–3.

87 "Successful Fish Advertising," *Saint John Telegraph-Journal*, 16 April 1937.

CHAPTER FOUR

1 See, for example, memo from W. Found, Assistant Deputy Minister, Department of Marine and Fisheries, 15 January 1921, General Information, Correspondence Regarding Encouraging the Production of Fish to Help the Shortage of Food, LAC, RG 23, vol. 509, folder 711-1-22 (5).

2 See General Information Brief, Minister's Tour of the Maritimes, 1929, Including Proceedings – Hearings, LAC, RG 23, vol. 511, folder 711-1-45 (1).

3 A.K. MacLean, Cyrus MacMillan, H.R.L. Bill, Joseph Mombourque, and J.G. Robichaud, *Report of the Royal Commission Investigating the Fisheries of the Maritime Provinces and the Magdalen Islands* (Ottawa: F.A. Acland, 1928), 5.

4 Ibid., 9.

5 Ibid.

6 Ibid., 32.

7 Ibid., 36–7.

8 A.F. Healy, MP, House of Commons, Windsor, Ont. to Ernest LaPointe, Minister of Marine and Fisheries, Ottawa, 19 July 1923, General Information, Advertising Campaign by the Canadian Co-operative Fish Publicity Fund to Increase Demand for Fish (hereafter Advertising Campaign by Co-operative Publicity Fund), July 1923 to July 1924, LAC, RG 23, vol. 528, folder 711-25-16 (1).

9 R. Sykes Muller, "Eat Fish Once a Day: Suggestions to Increase Consumption of Fish. Especially Prepared for Canadian Fisheries Association by R. Sykes Muller Co, Ltd., Montreal," 14 July 1923, "General Information, Advertising Campaign by Co-operative Publicity Fund, July 1923 to July 1924, LAC, RG 23, vol. 528, folder 711-25-16 (1).

10 Ibid.

11 William Knight, "Modeling Authority at the Canadian Fisheries Museum, 1884–1914" (PhD diss., Carleton University, Ottawa, 2014), 275–312.

12 Memo Re: Campaign to Increase Consumption of Fish in Canada, Signed J.J. Cowie (Acting Assistant Deputy Minister, Department of Marine and Fisheries, Fisheries Branch), 27 September 1923, General Information, Advertising Campaign by Co-operative Publicity Fund, July 1923 to July 1924, LAC, RG 23, vol. 528, folder 711-25-16 (1).

13 Committee of the Privy Council, PC 1999, General Information, by the Co-operative Publicity Fund, July 1923 to July 1924, LAC, RG 23, vol. 528, folder 711-25-16 (1). See also "$10,000 to Advertise Fish," *Yarmouth Telegram*, 26 October 1923.

14 B.T. Huston, Manager, Canadian Grocer, to J.A. Paulhus, President, Canadian Fisheries Association, 27 November 1923, General Information, Advertising Campaign by Co-operative Publicity Fund, July 1923 to July 1924, LAC, RG 23, vol. 528, folder 711-25-16 (1).

15 "Eat More Fish," advertisement for Educational Division, Canadian Fisheries Association, Montreal, *Canadian Grocer*, 19 January 1924, 11.

16 *Canadian Grocer* did publish some articles about individual stores and their spectacular fish sales, but there is little evidence that these cases represent a general pattern of increased consumption. "Sells One Mil-

lion Pounds Fish a Year," *Canadian Grocer*, 15 January 1924, 18; "Increase Sales of Canned Fish through Use of Window Display," *Canadian Grocer*, 15 January 1924, 18; "Sales of Fish Run Over 1,500 Pounds a Week," *Canadian Grocer*, 15 January 1924, 19; "Build Fish Trade in Quality," *Canadian Grocer*, 15 January 1924, 20.

17 J.H. Paulhus, President, Canadian Fisheries Association to Alexander Johnston, Deputy Minister of Marine and Fisheries, Ottawa, 2 June 1925, General Information, Advertising Campaign by Co-operative Publicity Fund, July 1923 to July 1924, LAC, RG 23, vol. 528, folder 711-25-16 (1).

18 "Fish Meals and Public Eating Places," *Canadian Fisherman*, September 1924, General Information, Advertising Campaign by Co-operative Publicity Fund, July 1923 to July 1924, LAC, RG 23, vol. 528, folder 711-25-16 (1).

19 "Help the Fisheries," *Halifax Herald*, 17 December 1925.

20 James H. Conlon, Chairman, Canadian Co-operative Fish Publicity Fund, Montreal to W.A. Found, Director of Fisheries, Ottawa, 7 February 1924, General Information, Advertising Campaign by Co-operative Publicity Fund, July 1923 to July 1924, LAC, RG 23, vol. 528, folder 711-25-16 (1).

21 J.H. Paulhus, President, Canadian Fisheries Association to Alexander Johnston, Deputy Minister of Marine and Fisheries, Ottawa, 2 June 1925, General Information, Advertising Campaign by Co-operative Publicity Fund, July 1923 to July 1924, LAC, RG 23, vol. 528, folder 711-25-16 (1).

22 Certified Copy of a Minute of a Meeting of the Committee of the Privy Council, 13 November 1925, General Information, Advertising Campaign by Co-operative Publicity Fund, n.d., LAC, RG 23, vol. 528, folder, 711-25-16 (3).

23 Cardin served as a MP for Richelieu from 1911 to 1946, first as a member of the ruling Liberal Party, then in Opposition, and finally as an Independent between 1945 and 1946. He served as the minister of marine and fisheries from 1924 to 1930 and then as the minister of marine after the duties were split into two ministries. He also served as the minister of public works from 1935 to 1942 and minister of transportation from 1940 to 1942.

24 G.H. Langtry, Secretary, Nova Scotia Sea Fisheries Association, Yarmouth, Nova Scotia to P.J.A. Cardin, Minister of Marine and Fisheries, Ottawa, 4 December 1925, General Information, Advertising Campaign by Co-operative Publicity Fund, n.d., LAC, RG 23, vol. 528, folder, 711-25-16 (3).

25 Telegram from Alexander Johnston, Deputy Minister of Marine and Fisheries, to Arthur Boitelier, President of Canadian Fisheries Association, 7 December 1925, General Information, Advertising Campaign by

Co-operative Publicity Fund, n.d., LAC, RG 23, vol. 528, folder, 711-25-16 (3).

26 The exchanges of telegraph communications can be found in General Information, Advertising Campaign by Co-operative Publicity Fund, n.d., LAC, RG 23, vol. 528, folder, 711-25-16 (3).

27 F.T. James, F.T. James Company, Toronto, to BK Snow, Chairman, Publicity Committee, Canadian Fisheries Association, Halifax, General Information, Advertising Campaign by Co-operative Publicity Fund, n.d., LAC, 19 March 1926, RG 23, vol. 528, folder, 711-25-16 (3).

28 B.K. Snow, Chairman, Publicity Committee, Canadian Fisheries Association, Halifax, to F.T. James, F.T. James Company, Toronto, 19 March 1926, General Information, Advertising Campaign by Co-operative Publicity Fund, n.d., LAC, RG 23, vol. 528, folder 711-25-16 (3).

29 Cowie, "Memo, Re: Renewal of the Fish Advertising Campaign," 15 October 1926, General Information, Advertising Campaign by Co-operative Publicity Fund, n.d., LAC, RG 23, vol. 528, folder 711-25-16 (3).

30 Wm. A. Found, for Deputy Minister of Marine and Fisheries to J.T. O'Connor, Vice President, Canadian Fisheries Association, Montreal, 7 October 1926, General Information, Advertising Campaign by Co-operative Publicity Fund, n.d., LAC, RG 23, vol. 528, folder 711-25-16 (3).

31 Original French: "Un tel octori des fonds publices'est d'aucune utilité pour les pêcheurs." Translations provided within the documentation by government translator. Hilaire Samson, President, and Andrew Sampson, Secretary, Local 2 of Fishermen's Federation of Nova Scotia, Petit Degrat (Richmond), Nova Scotia, to P.J.A. Cardin, Minister of Marine and Fisheries, Ottawa, 30 January 1928, General Information, Advertising Campaign by Co-operative Publicity Fund, n.d., LAC, RG 23, vol. 528, folder 711-25-16 (3).

32 Original French: "Les chalutiers ne font du bien qu'aux détenteurs d'actions dans les Compagnies, une poignée d'hommes assez riches, comparé evolut à de milliers de pêcheurs qui, avec leur famille, souffrent et sont même forces de quitter le pays, á cause de ces chalutiers."

33 Original French: "l'ennemi mortel du pêcheur."

34 Robb Robinson, *Trawling: The Rise and Fall of the British Trawl Fishery* (Exeter: University of Exeter Press, 1996), 15–23.

35 Ibid., 90–112.

36 Ibid., 24–38.

37 Ibid., 50–3, 98.

38 Matthew McKenzie, *Breaking the Banks: Representation and Realities in*

New England Fisheries, 1866–1966 (Amherst: University of Massachusetts Press, 2018), 37–55.

39 MacLean et al., *Report of the Royal Commission*, 91–2.

40 Ibid., 95.

41 Ibid., 98.

42 The report actually argues that shore fishermen should increase their productivity. Ibid., 100.

43 Ibid., 102–15.

44 "Report of Fishermen's Meeting Held in Phalen Hall (Canso), 15 July 1927, Expansion of Fish Trade, Development of Fisheries in Nova Scotia, LAC, RG 23, vol. 528, folder 711-25-17 (1). Fishermen's unions also submitted numerous petitions to the minister of marine and fisheries during his regional tour in 1929; see Part I: General Information Brief, Minister's Tour of the Maritimes, 1929, Including Proceedings – Hearings, LAC, RG 23, vol. 511, folder 711-1-45.

45 Extract from meeting of Maritime Fishermen in Halifax, NS, during January 1931, Advertising Campaign, Requested by Canadian Fisheries Association, Generally, February 1931 to 16 July 1931, LAC, RG 23, vol. 531, folder 711-25-19 (3).

46 Canada grocery trade papers noted the rising popularity of fresh fish; see "The New and Easy Way to Sell Fish," advertising for Austin E. Nickerson, Limited, Yarmouth, NS, *Canadian Grocer*, 3 August 1928, 41. The National Fish Company was a regular advertiser in the *Canadian Grocer* and often used that advertising to market the superiority of fresh fish from its trawler fleet. See "The Pickled Period Is Past," advertisement for National Fish Co. Ltd, The King Fishers of Canada, Owners of the National Fleet of Trawlers and Auxiliary Vessels, Largest Curers of Smoked Fish in America. Halifax and Port Hawkesbury, *Canadian Grocer*, 3 August 1928, 42. In this ad, the National Fish Company claimed to have "revolutionised" the seafood industry in Canada and to be the "pioneer" in the fresh fish trade.

47 Brief by Maritime-National Fish Limited, Halifax, NS, presented at the Fisheries Conference, 13–14 July 1938, Expansion of Fishing Trade, Conference Re Fisheries in Nova Scotia, June 1938 to 11 August 1938, LAC, RG 23, vol. 540, folder 711-25-28 (1).

48 Memorandum Submitted at Fisheries Conference, Halifax, Nova Scotia, Station #101 of the Fishermen's Federation of Nova Scotia, Expansion of Fishing Trade, Conference Re Fisheries in Nova Scotia, June 1938 to 11 August 1938, LAC, RG 23, vol. 540, folder 711-25-28 (1).

49 Coverage of the CFA lobbying was carried by many interested media

outlets such as in "Advertising Our Fish," *Saint John Telegraph Journal*, 13 July 1928, and "Would Spend $100,000 to Advertise Fish Food," *Halifax Herald*, 10 July 1928.

50 This 1928 resolution was quoted in Memorandum to the Government Respecting Publicity Campaign for the Fishing Industry and Tax Imposed on Trawlers, Canadian Fisheries Association Executive Committee, Montreal, 10 May 1930, Advertising Campaign, Requested by Canadian Fisheries Association, 2 August 1929 to 24 September 1930, LAC, RG 29, vol. 531, folder 711-25-19 (1).

51 "Royal Commission Probe of Fishing Conditions Is Urged by Canso Meeting," *Halifax Chronicle*, 15 July 1927; "Save the Fishermen, Prices So Low That the Fish Are Not Worth Catching," *Halifax Chronicle*, 16 July 1927.

52 "B.C. Salmon to Be Advertised, Govt. Contributes to Fund Provided by Canners," *Vancouver Sun*, 5 June 1930; "Tolmie Will Help to Advertise Salmon," *Vancouver Daily Province*, 12 August 1930.

53 McConnell & Fergusson, Ltd, Advertising Agency, Vancouver, Canada, to Canned Salmon Advertising Committee, Vancouver, BC, 31 October 1930, Advertising Campaign, Requested by Canadian Fisheries Association, Generally, September 1930 to 30 January 1931, LAC, RG 23, vol. 531, folder 711-25-19 (2).

54 Stock media language provided by McConnell & Fergusson, Ltd, March 1931, Advertising Campaign, Requested by Canadian Fisheries Association, Generally, February 1931 to 16 July 1931, LAC, RG 23, vol. 531, folder 711-25-19 (3).

55 "This Is Action," *Vancouver Sun*, 16 December 1930. See also "Department to Aid Fish Advertising," *Saint John Telegraph Journal*, 16 December 1930; "To Advertise Fish," *Halifax Chronicle*, 16 December 1930; "Help Sale of Salmon," *Daily News* (Prince Rupert), 16 December 1930; "Helping to Sell Salmon," *Gazette* (Montreal), 19 December 1930; "Seek Dominion Aid in Advertising Fish," *Vancouver Daily Province*, 18 December 1930.

56 "Salmon Industry Romance Unfolded at Banquet Here," *Ottawa Evening Journal*, 13 January 1931; "Eat More Salmon, Plea of Nation-Wide Campaign," *Vancouver Daily Province*, 13 January 1931; "Salmon Men Urged to Promote Sales," *Vancouver Sun*, 13 January 1931; "Protecting the Salmon Industry," *Ottawa Journal*, 14 January 1931; "Buy Our Own Salmon," *Daily News* (Prince Rupert), 16 January 1931; "To Inaugurate B.C. Salmon Advertising Campaign," *Casco Breeze* (Truro, NS), 17 January 1931; "To Spend $50,000 Advertising Salmon," *Ottawa Evening Jour-

nal, 19 January 1931; "Salmon Fishing One of Canada's Largest Indus-
tries, Says Gosse, in Giving Interview in Montreal," *Daily News* (Prince
Rupert), 20 January 1930; "Campaign for Market in Canada," *Halifax
Herald*, 21 January 1931; "Will Inaugurate Fish Campaign," *Halifax Her-
ald*, 21 January 1931; "Canned Salmon Is Safe Food," *Halifax Herald*, 29
January 1931.

57 P.J. Cardin, Minister of Fisheries, to Hon. J. Malcolm, Minister of Trade
and Commerce, Ottawa, 30 April 1930, Advertising Campaign, Requested by Canadian Fisheries Association, 2 August 1929 to 4 September
1930, LAC, RG 29, vol. 531, folder 711-25-19 (1).

58 A.N. McLean, President, Connors Brothers Ltd, Saint John, NB, to Wm.
A. Found, Deputy Minister of Fisheries, 10 June 1930, and Wm. A.
Found, Deputy Minister of Fisheries, to A.N. McLean, President, Connors Brothers Ltd, Saint John, NB, 13 June 1930, Advertising Campaign,
Requested by Canadian Fisheries Association, 2 August 1929 to 24 September 1930, LAC, RG 29, vol. 531, folder 711-25-19 (1). Regional newspapers helped promote McLean's concerns of industry favouritism. See,
for example, "Increased Markets," *Fundy Fishermen* (Saint John), 6 February 1931.

59 A.H. Brittain, "Report of the Chairman of Publicity Committee, Canadian Fisheries Association, at the Annual General Convention," Prince
Rupert, BC, 2 August 1929, Advertising Campaign, Requested by Canadian Fisheries Association, 2 August 1929 to 24 September 1930, LAC,
RG 29, vol. 531, folder 711-25-19 (1). Some of this was covered in the
newspapers; for example, "Seeks $100,000 Fish Publicity," *Vancouver
Daily Province*, 2 August 1929; "$100,000 for Publicity Asked for Fish-
eries," *Globe* (Toronto), 16 September 1929; "Fish Interests Seek $100,000
for Publicity," *Halifax Herald*, 19 September 1929.

60 A.H. Brittain, "Report of the Chairman of Publicity Committee, Canadian Fisheries Association, at the Annual General Convention," Prince
Rupert, BC, 2 August 1929, Advertising Campaign, Requested by Canadian Fisheries Association, 2 August 1929 to 24 September 1930, LAC,
RG 29, vol. 531, folder 711-25-19 (1).

61 Memorandum to the Government Respecting Publicity Campaign for
the Fishing Industry and Tax Imposed on Trawlers," Canadian Fisheries
Association Executive Committee, Montreal, 10 May 1930, Advertising
Campaign, Requested by Canadian Fisheries Association, 2 August 1929
to 24 September 1930, LAC, RG 29, vol. 531, folder 711-25-19 (1).

62 Burnaby had a long history in commerce. In the 1920s, he was president
of the United Farmers of Ontario and was briefly caught up in a bribery

scandal. He said he had been offered a bribe but denied taking it, and no evidence exists that he did. Burnaby was active in farmers' politics (even trying to start a third party) and public relations for his opposition to tariffs as part of his work with the United Farmers of Ontario. See, for example, *Ottawa Citizen*, 30 April 1921; "Mr. Burnaby at Chamber," *Windsor Star*, 14 May 1931; "Maritime Trade," *Windsor Star*, 15 May 1931; "Deaths," *Windsor Star*, 31 October 1959.

63 Burnaby's speech was widely covered in the trade-friendly and regional press. See "Seek Dominion Aid in Advertising Fish," *Vancouver Daily Province*, 18 December 1930; "To Establish Fishing Industry on Sound Basis," *Charlottetown Guardian*, 18 December 1930; "Burnaby Urges Fish Advertising," *Saint John Telegraph Journal*, 18 December 1930; "Fishing Need," *Halifax Herald*, 18 December 1930; "Selling More Fish," *Saint John Telegraph Journal*, 19 December 1930; "Boost Food Value of Fish by Ads," *Evening Patriot* (Charlottetown), 19 December 1930; "Advertising Fish," *Saint John Telegraph Journal*, 20 December 1930; "Urges Advertising of Maritime Fish," *Saint John Telegraph Journal*, 23 December 1930; "Extensive Advertising of Fish Recommended," *Fundy Fisherman*, 26 December 1930; "Necessity Promoting Canadian Sales Stressed. Foreign Field to Be Pushed," *Halifax Chronicle*, 13 January 1931.

64 See, for example, a number of telegrams exchanged between R.W. Gould, Secretary, Canadian Fisheries Association, and William Found, Deputy Minister, Department of Fisheries, and P.J.A. Cardin, Minister of Fisheries, as well as resolutions passed by the Canadian Fisheries Association at their annual meetings calling for government-funded advertising, all of which were forwarded to the Ministry of Fisheries, found in Advertising Campaign, Requested by Canadian Fisheries Association, 2 August 1929 to 24 September 1930, LAC, RG 29, vol. 531, folder 711-25-19 (1).

65 Wm. A. Found, Deputy Minister of Fisheries, to the Manager, Canadian National Railway Broadcasting Station, Ottawa, 8 October 1929, Development of Fisheries, Advertising Fisheries through Radio Broadcasting, LAC, RG 23, vol. 528, folder 711-25-18 (1).

66 "The Canadian Fisheries and the Work of the Department of Fisheries," by Wm. A. Found, D.Sc., Deputy Minister of Fisheries, for Broadcast under the auspices of the Professional Institute of the Civil Service of Canada, Wednesday, 24 February 1932, CNRO, Ottawa, 6–7, Information, Radio Broadcast, 25 June 1930 to 7 February 1934, LAC, RG 23, vol. 528, folder 711-25-18 (2).

67 Minutes of a Meeting of the Committee of the Privy Council, PC 2595, 7 November 1930.

68 "Progress Report on the Survey of Markets and Marketing Methods for Canadian Fish and Fish Products," to the Honourable E.N. Rhodes, Minister of Fisheries, prepared by Cockfield, Brown, & Company Ltd, Montreal, Toronto, Winnipeg, Vancouver, March 1931, Expansion of the Fishing Trade, Survey of Fishing Industry, 7 November 1930 to 25 November 1931, LAC RG 23, vol. 531, folder 711-25-22 (1).

69 Cockfield, Brown & Company Ltd, Montreal, Toronto, Winnipeg, Vancouver, "Summary of Report on the Marketing of Canadian Fish and Fish Products, to the Honourable E.N. Rhodes, Minister of Fisheries," 1931, 1–10, Expansion of Fish Trade, Survey of Fishing Industry, 19 January 1932, LAC, RG 23, vol. 531, folder 711-25-22 (3).

70 Ibid., 11–16.

71 Ibid., 121.

72 Ibid., 19–22.

73 Ibid., 30.

74 Ibid., 66.

75 Ibid, 100–4.

76 Ibid., 122.

77 Ibid.,161–2.

78 This information and what follows is largely drawn from John Herd Thompson and Allen Seager, *Canada, 1922–1939: Decades of Discord* (Toronto: McClelland & Stewart, 1985); H. Blair Neatby, *The Politics of Chaos: Canada in the Thirties* (Toronto: Macmillan of Canada, 1972). See also Denyse Baillargeon, *Making Do: Women Family and Home in Montreal during the Great Depression*, translated by Yvonne Klein (Waterloo: Wilfrid Laurier University Press, 1999), Lara Campbell, *Respectable Citizens: Gender, Family, and Unemployment in Ontario's Great Depression* (Toronto: University of Toronto Press, 2009); James Struthers, *The Limits of Affluence: Welfare in Ontario, 1920–1970* (Toronto: University of Toronto Press, 1994).

79 Resolution, 13 February 1934, House of Commons Committees, 17th Parliament, 5th Session: *Special Committee on Price Spreads and Mass Buying*, vol. 1 (Ottawa: J.O. Patenaude), iii. See also in General Information, Information Re. Inquiry into the Spread of Prices of Fish, General, February 1934, LAC, RG 23, vol. 511, folder 711-1-51 (1).

80 L.W. Fraser, "Report of Enquiry into the Fishing Industry, Submitted on Behalf of the Price Spreads and Mass Buying Committee," Appendix A, 3904–25, in General Information, Information Re. Inquiry into the Spread of Prices of Fish, General, February 1934, LAC, RG 23, vol. 511, folder 711-1-51 (1).

81 Ibid., Appendix A, 3908, 3909.

82 Ibid., Appendix A, 3909, 3922. ·

83 Fraser's presentation of his report did not so neatly fall into these two categories; this is just an interpretation of the language presented in the report.

84 Fraser, "Report of Enquiry into the Fishing Industry," Appendix A, 3922–3.

85 Ibid., 3909, 3923.

86 J.J. Cowie, Internal Memorandum for Wm. A. Found, Department of Fisheries, General Information, Information Re. Inquiry into the Spread of Prices of Fish, General, February 1934, LAC, RG 23, vol. 511, folder 711-1-51 (1).

87 "The Price Spreads Reports, Several Factors Ignored in Recommendations," *Canadian Fisherman* 22, no. 5 (May 1935): 1.

88 Ibid., 2.

89 "Sees Market for Fish, A.H. Brittain Says Advertising Increases Consumption," *Gazette* (Montreal), 5 November 1935.

90 J.A. Paulhus, President, Canadian Fisheries Association, "Industry Needs Publicity Campaign to Enlarge Home Market, Better Times Looming Ahead," *Canadian Fisherman*, October 1933, 11–12.

91 S.M. Gould, Secretary Treasurer, Canadian Fisheries Association, Montreal, to Hon. Alfred Duranleau, Acting Minister of Fisheries, Ottawa, 16 October 1933, Expansion of Fish Trade, Generally, 26 September 1933 to 26 June 1936, RG 23, vol. 532, folder 711-25-24 (1).

92 "Campaign Promises Fisheries Aid," *Gazette* (Montreal), 24 September 1935; "Duff Snubbed by Fishermen," *Charlottetown Guardian*, 24 September 1935.

93 See, for example, "Bonus Plan Is Urged," *Halifax Herald*, 24 September 1935; "Fisheries Assn. Urges Publicity Campaign," *Charlottetown Patriot*, 25 September 1935; "Fishermen Entitled to Adequate Living," *Halifax Herald*, 25 September 1935; "Aid Promised Fishermen, Rehabilitation of Markets Will Be Discussed, Bennett Says," *Globe* (Toronto), 26 September 1935; "Fisheries Request No Political Move, Capt. Wallace Insists Bennett Should Take Pre-Election Stand," *Gazette* (Montreal), 7 October 1935.

94 "Says Shore Fishermen Received No 'Santa Claus Visits,' Statement Criticizes Proposals," *Halifax Chronicle*, 9 October 1935; William P. Groom, Letter to Editor, "Is Opposed to Proposals of Organization," *Halifax Herald*, 10 October 1935.

95 J.J. Kinley, MP, House of Commons, Lunenburg, Nova Scotia, to Wm. A. Found, Deputy Minister of Fisheries, Ottawa, 25 November 1939, General Information, Information Re: Enquiry into Spread in Price of Fish, 10 September 1934 to 30 November 1936, LAC, RG 23, vol. 511, folder 711-1-51 (2).

96 Scores of newspaper clippings discussing the flight of fishermen can be found in General Information, Information Re: Enquiry into Spread in Price of Fish, 7 December 1937 to 25 August 1938, LAC, RG 23, vol. 511, folder 711-1-51 (3).

97 Hon. J.E. Michaud, Minister of Fisheries, Ottawa, to R.W. Widdess, President, Seaport Crown Fish Company, Vancouver, BC, 20 November 1935, Expansion of Fish Trade, Generally, 26 September 1933 to 26 June 1936, RG 23, vol. 532, folder 711-25-24 (1).

98 Frederick Wallace, Canadian Fisheries Association, General Memorandum Re Canadian Fishing Industry, to Hon, J.E, Michaud, Minister of Fisheries, Montreal, 13 December 1935, Expansion of Fish Trade, Generally, 26 September 1933 to 26 June 1936, LAC, RG 23, vol. 532, folder 711-25-24 (1).

99 Ibid.

100 The Fisheries Committee of the Halifax Board of Trade on behalf of the Fresh Fish Industry of Nova Scotia, Submitted to Hon. J.B. Michaud, Minister of Fisheries, 29 January 1936. See also Resolution of the Lunenburg Board of Trade, 18 February 1936, Expansion of Fish Trade, Generally, 26 September 1933 to 26 June 1936, LAC, RG 23, vol. 532, folder 711-25-24 (1).

101 Minute of a Meeting of the Committee of the Privy Council, PC 1804, 17 July 1936, Expansion of Fish Trade, Generally, 27 June 1936 to 7 August 1936, LAC, RG 23, vol. 532, folder 711-25-24 (2). This amount would be roughly equivalent to $4.024 million in 2022 dollars, as calculated by Bank of Canada's Inflation Calculator, http://www.bankofcanada.ca/rates/related/inflation-calculator/.

102 For more details on the expenditures, see "Report of E.W. Reynold & Co. Ltd to Honourable J.E. Michaud, Department of Fisheries, Ottawa, 10 February 1939," Advertising Fisheries, Generally, 7 February 1939 to 28 February 1939, LAC, RG 23, vol. 536, folder 711-25-24 (31).

103 "Dominion Government to Promote Consumption of Fish, Citizens to Be Made More Aware of the Treasures of Health, Energy and Enjoyment Abounding in Canada's Seas and Lakes," form language produced by E.W. Reynolds & Co. Ltd for circulation to Canadian newspa-

pers, 28 July 1936, Expansion of Fish Trade, Generally, 27 June 1936 to 7 August 1936, LAC, RG 23, vol. 532, folder 711-25-24 (2).

104 "Put Down Your Bucket Where You Are!," publication sent by Atlantic Advertising Agency to all teachers in Nova Scotia, Prince Edward Island, and New Brunswick, Expansion of Fishing Trade, 23 November 1936 to 16 December 1936, LAC, RG 23, vol. 533, folder 711-25-24 (10).

105 Atlantic Advertising Agency, 23 January 1937, Radio Spots, Expansion of Fishing Trade, 12 January 1937 to 30 January 1937, LAC, RG 23, vol. 533, folder 711-25-24 (12).

106 Radio Spots by Atlantic Advertising Agency, 4 February 1937, Expansion of Fishing Trade, 1 February 1937 to 13 February 1937, LAC, RG 23, vol. 533, folder 711-25-24 (13).

107 Editorial, *Charlottetown Guardian*, 6 February 1936.

108 "Fish Marketing Campaign to Continue," *Canadian Fisherman* 25, no. 6 (June 1938): 1.

109 "Advertising Helps This Basic Industry," *Ottawa Morning Journal*, 4 February 1937.

110 "Report of E.W. Reynold & Co. Ltd to Honourable J.E. Michaud, Department of Fisheries, Ottawa, February 10, 1939," Advertising Fisheries, Generally, 7 February 1939 to 28 February 1939, LAC, RG 23, vol. 536, folder 711-25-24 (31).

111 C.J. Morrow, Lunenburg Sea Products, Lunenburg, NS, to W.A. Found, Deputy Minister of Fisheries, Ottawa, 3 November 1936, Expansion of Fishing Trade, 28 October 1936 to 6 November 1936, LAC, RG 23, vol. 533, folder 711-25-24 (8).

112 "Fish Buyers Should Enquire about Country of Origin," *Maritime Merchant*, 4 May 1939, 15.

113 "Estimated 25% Increase in Fish Sales as a Result of Advertising Campaign, Competition of Cheap Meat Largely Offset – Many New Retail Outlets Opened Following Advertising That Seeks to Make Canada 'Fish Conscious,'" *Marketing* (Toronto) 42, no. 5 (January 1937).

114 Report of E.W. Reynold & Co. Limited to Honourable J.E. Michaud, Department of Fisheries, Ottawa, 10 February 1939, Advertising Fisheries, Generally, 7 February 1939 to 28 February 1939, LAC, RG 23, vol. 536, folder 711-25-24 (31).

115 George Akins, Vice President, Walsh Advertising, to E.L. Michaud, Minister, Department of Fisheries, 20 May 1940, Advertising Fisheries, Generally, 22 April 1940 to 24 June 1940, LAC, RG 23, vol. 536, folder 711-25-24 (40).

116 The firm's 1939 report stated that it had reached 937 publications in 1938–39, totalling 37,049,422 advertising messages, and distributed 250,000 fish cookbooks. See Report of E.W. Reynold & Co. Limited to Honourable J.E. Michaud, Department of Fisheries, Ottawa, 10 February 1939, Advertising Fisheries, Generally, 7 February 1939 to 28 February 1939, LAC, RG 23, vol. 536, folder 711-25-24 (31).

117 A.N. McLean, Connors Brothers Ltd, Saint John, NB, to Canadian Fisheries Association, Montreal, 8 March 1937, Expansion of Fishing Trade, 10 March 1937 to 23 March 1937, LAC, RG 23, vol. 534, folder 711-25-24 (15).

118 William Found, Deputy Minister, Department of Fisheries, letter to the editor, *The World* (Cobourg, ON), 19 December 1938, Advertising the Fisheries, Generally, 27 October 1938 to 19 October 1938, LAC, RG 23, vol. 535, folder 711-25-38 (29).

119 "Successful Fish Advertising," *Saint John Telegraph-Journal*, 16 April 1937.

120 As reported in "Fish Business Shows Revival Minister Says, Output Greatest in Value in 6 Years – Advertising Campaign Brings Results," *Evening Telegram* (Toronto), 13 September 1937. See also "Importance of Wider and More Diverse Markets Emphasized at Recent Annual Meeting of the Canadian Fisheries Association – Progress Continues," *Evening Empire* (Prince Rupert), 21 October 1937; "For More Fish," *Ottawa Journal*, 29 October 1937.

121 R.W. Gould, Secretary-Treasurer, Canadian Fisheries Association, Montreal, to Hon. J.E. Michaud, Minister of Fisheries, Ottawa, 11 May 1937, Expansion of Fishing Trade, Advertising, General, 4 May 1937 to 7 September 1937, LAC, RG 23, vol. 534, folder 711-25-24 (17).

122 "Fish Business Shows Revival Minister Says, Output Greatest in Value in 6 Years – Advertising Campaign Brings Results," *Evening Telegram* (Toronto), 13 September 1937.

123 Minute of a Meeting of the Committee of the Privy Council, Approved by His Excellency the Governor General on the 6 October 1937, Expansion of Fishing Trade, Advertising, 8 September 1937 to 18 October 1937, LAC, RG 23, vol. 534, folder 711-25-24 (18).

124 Minute of a Meeting of the Committee of the Privy Council, PC16, 8 January 1837, Expansion of Fishing Trade, 17 December 1936 to 11 January 1937, LAC, RG 23, vol. 533, folder 711-25-24 (11). See also William Found, Deputy Minister, Department of Fisheries, to R.E. Reed, Department of Sea and Shore Fisheries, Maine, 28 April 1937, Expansion of Fishing Trade, 24 March 1937 to 1 May 1937, LAC, RG 23,

vol. 534, folder 711-25-24 (16). These funds were later augmented with an additional $7,000, bringing the total advertising funding for October 1937 to March 1938 to $107,000. This is roughly equivalent to $2.042 million in 2022 dollars; see http://www.bankofcanada.ca/rates/related/inflation-calculator/.

125 R.W. Gould to Hon. J.E. Michaud, Minister of Fisheries, 11 April 1938; includes Resolution Passed at Meeting of the Board of Directors of the Canadian Fisheries Association, Tuesday, 5 April 1938, Advertising the Fisheries, Generally, 12 March 1938 to 22 April 1938, LAC, RG 23, vol. 535, folder 711-25-38 (25).

126 R.W. Gould to Hon. J.E. Michaud, Minister of Fisheries, 19 May 1938, Advertising Fisheries, Generally, 25 April 1938 to 28 June 1938, LAC, RG 23, vol. 535, folder 711-25-38 (26).

127 "Sea Fisheries Conference, Bay of Fundy Coast, New Brunswick, 23 August 1938," Expansion of Fishing Trade, Conference on New Brunswick Fisheries, 14 September 1938 to 16 September 1938, LAC, RG 23, vol. 540, folder 711-25-29 (2). For press coverage of Michaud's speech, see "Sea Fisheries Conference, Bay of Fundy Coast, New Brunswick," 23 August 1938; "New Brunswick Fisheries Ills Aired," *Saint John Telegraph-Journal*, 23 August 1938; "Results 'More Than Please' States Hon. J.E. Michaud," *Saint John Telegraph-Journal*, 24 August 1938; "Fish Conferences Are Held in New Brunswick," *Fundy Fishermen* (Black's Harbour, NB), 24 August 1938.

128 Minute of a Meeting of the Committee of the Privy Council, approved by His Excellency the Administrator on 4 August 1938, PC 1778, Advertising the Fisheries, Generally, 29 June 1938 to 8 September 1938, LAC, RG 23, vol. 535, folder 711-25-38 (27). See also "Fish Marketing Campaign to Continue," *Canadian Fisherman* 25, no. 6 (June 1938): 1. This is roughly equivalent to $2.546 million in 2022 dollars; see http://www.bankofcanada.ca/rates/related/inflation-calculator/.

129 Gould, Secretary Treasure, Canadian Fisheries Association to Michaud, 27 April 1939 including resolutions of the board meeting 27 April 1939, Advertising Fisheries, Generally, 5 April 1939 to 21 June 1939, LAC, RG 23, vol. 536, folder 711-25-24 (33).

130 Minute of a Meeting of the Committee of the Privy Council, approved by the Deputy of his Excellency the Governor General on the 8 June 1939, PC 1404, Advertising Fisheries, Generally, 5 April 1939 to 21 June 1939, LAC, RG 23, vol. 536, folder 711-25-24 (33).

131 W. George Akins, Walsh Advertising Company Ltd, Toronto, Memorandum: Department of Fisheries, Ottawa, 12 September 1939, Adver-

tising Fisheries, Generally, August 1939 to 26 September 1939, LAC, RG 23, vol. 536, folder 711-25-24 (35).

132 "Britain Stopped Lobster Imports," *Canadian Grocer*, 15 September 1939, 8.

133 Privy Council, PC 2283, 30 May 1940, provided $50,000 to fund the advertisement of Canada Brand Lobster, Advertising Fisheries, Generally, 22 April 1940 to 24 June 1940, LAC, RG 23, vol. 536, folder 711-25-24 (40).

134 W. George Akins, "A Report on the Advertising Campaign to Promote the Sale of Canadian Canned Lobster," 31 October 1940, Advertising Fisheries, Generally, 24 October 1940 to 31 January 1941, LAC, RG 23, vol. 538, folder 711-25-24 (44).

135 Privy Council, PC 378, 29 January 1940, Advertising Fisheries, Generally, 20 December 1939 to 21 February 1940, LAC, RG 23, vol. 537, folder 711-25-24 (38).

136 "Be Patriotic! Serve Lobster, It's a Pleasant Way to Serve Canada," *Gazette* (Montreal), Monday, 22 July 1940, 5; "Women of Canada, Your Help Is Needed," advertisement for Department of Fisheries, *Canadian Home Journal*, June 1941, 55.

137 Text of Radio Talk by Monica Mugan, Radio Station C.B.L. & C.B.Y. Toronto, CBC, 18 September 1940, Advertising Fisheries, Generally, 17 September 1940 to 23 October 1940, LAC, RG 23, vol. 536, folder 711-25-24 (43).

138 "Retail Sales of Canned Lobster Are Stimulated by Government Advertising," *Retail Grocer and Provisioner*, November 1940, 36; "Lobsters Go to New Markets," *Toronto Board of Trade Journal* 30, no. 11 (November 1940): 3, 15; George Akins, Vice President, Walsh Advertising, to L.E. Michaud, Minister of Fisheries, 1 April 1942, Advertising Fisheries, Generally, 16 September 1941 to 25 June 1942, LAC, RG 23, vol. 538, folder 711-25-24 (48).

139 "More Fish Included in Army Rations," *Retail Grocer and Provisioner*, November 1940, 36.

140 "Vigorous Government Promotion Saves War-Hit Lobster Industry," *Marketing*, 23 November 1940.

141 "Immediate Abolition of Beam Trawler Demanded," *Sherbrooke Daily Record*, 16 April 1935. This article includes fishermen's protest against the beam trawler; it thanked none other than Leonard W. Fraser. Fraser had written the fisheries report for the 1934 Price Spread Commission, in which he called upon the industry and government to focus on quality of production, not quantity, and warned that efforts to market

fish would fail unless quality of product was first addressed. Although the Department of Fisheries dismissed and ignored Fraser's report, Fraser went on to represent Cumberland County in the Nova Scotia House of the Assembly between 1940 and 1941 and was leader of the opposition as a member of the Progressive Conservative Party. Fraser appears to have continually voiced his support of shore fishermen in their campaign against the beam trawler.

142 J.J. Kinley, MP, Lunenburg, Nova Scotia, to W.A. Found, Deputy Minister, Department of Fisheries, 5 November 1936, Expansion of Fishing Trade, 7 November 1936 to 21 November 1936, LAC, RG 23, vol. 533, folder 711-25-24 (9).

143 W. George Akins, Walsh Advertising Co. Ltd, Toronto, to Hon. J.E. Michaud, Minister of Fisheries, 20 June 1940, Advertising Fisheries, Generally, 22 April 1940 to 24 June 1940, LAC, RG 23, vol. 536, folder 711-25-24 (40).

144 C.J. Murrow, Lunenburg Sea Products, Lunenburg, Nova Scotia, to Dr. D.B. Finn, Deputy Minister, Department of Fisheries, 14 June 1940, Advertising Fisheries, Generally, 22 April 1940 to 24 June 1940, LAC, RG 23, vol. 536, folder 711-25-24 (40).

145 Stewart Bates, Professor of Commerce, Dalhousie University, Halifax, Nova Scotia, "The Canadian Atlantic Sea Fisheries," report written on behalf of the Provincial Government in accordance with arrangements made between Premier of Nova Scotia and Minister of Fisheries, Ernest Bertrand, General Information, Expansion of Fishing Trade, Report on the Nova Scotia Fishing Industry, by Stewart Bates, 27 August 1943 to 30 June 1944, LAC, RG 23, vol. 540, folder 711-25-30 (1).

146 Bates, "Canadian Atlantic Sea Fisheries," 9.

147 "Forty Fathoms," No. 9 of Conserving Canada Radio Broadcasts of CBC, 1 December 1944, General Information, Radio Broadcasts, 17 October 1938 to 25 February 1946, LAC, RG 23, vol. 528, folder 711-25-18 (5). In 1949, Richard S. Lambert would win the first Governor General's Award for Juvenile Fiction for his *Franklin of the Arctic: A Life of Adventure.*

148 H.F.S. Paisley, Director of Fisheries, to R.S. Lambert, Esq., Programme Director, Canadian Broadcasting Corporation, Toronto, 22 June 11944, General Information, Radio Broadcasts, 17 October 1938 to 25 February 1946, LAC, RG 23, vol. 528, folder 711-25-18 (5).

149 Barry joined the Department of Marine and Fisheries in 1924 as a district inspector in the Maritimes. He became district supervisor at Newcastle, New Brunswick, in 1928 and was director of eastern fisheries in

the Department of Fisheries until that ministry's reorganization. He became chief supervisor in Halifax and director of eastern fisheries in 1940. He was awarded the Military Cross for his service in the First World War and remained honorary colonel of the North Shore (NB) Regiment. See *Trade News*, March 1951, 20.

150 Telegram from H.F.S. Paisley to Colonel A.L. Barry, Chief Supervisor of Fisheries, Halifax, 8 September 1944, General Information, Radio Broadcasts, 17 October 1938 to 25 February 1946, LAC, RG 23, vol. 528, folder 711-25-18 (5).

151 "Memorandum Re Activities 1940–1952 Department of Fisheries – Fisheries Research Board," 22 August 1952, Information, Expansion and Development of the Fisheries, October 1949 to 16 December 1952, LAC, RG 23, vol. 541, folder 711-23-36 (3).

CHAPTER FIVE

1 Propaganda was certainly not the only means the Department of Fisheries used to encourage growth in the fisheries industry or to contribute to the war effort. It also subsidized fishing vessel construction, worked with the British Ministry of Food to provide canned salmon, canned herring, and frozen fillets, and worked with the Combined Food Board and the Wartime Price and Trade Board. Many of these efforts continued into the postwar relief and rehabilitation programs. See Memorandum Re Activities 1940–1952, Department of Fisheries – Fisheries Research Board, 22 August 1952, Information, Expansion and Development of the Fisheries, October 1949 to 16 December 1952, LAC, RG 23, vol. 541, folder 711-23-36 (3).

2 "Fish Becomes Increasingly Important; Minister U[r]ges Co-operation by Trade," *Canadian Grocer*, 1 November 1939, 24.

3 "Patriotic Citizens Aid Britain by Eating Fish for Breakfast," *Gazette* (Montreal) 14 July 1941, 6; "Fish Urged as Ideal Breakfast Now That Bacon Sale Is Curbed," *Gazette* (Montreal), 28 June 1941; "Eat More Fish," *Hamilton Spectator*, 18 July 1941.

4 Newspaper copy language included in J.M. Bowman, Walsh Advertising Company, to H.F.S. Paisley, Director of Publicity, Department of Fisheries, 23 September 1939, Advertising Fisheries, Generally, August 1939 to 26 September 1939, LAC, RG 23, vol. 536, folder 711-25-24 (35).

5 The language was initially provided in W. George Akins, Walsh Advertising Company Limited, Toronto, memorandum: Department of Fisheries, Ottawa, 12 September 1939, Advertising Fisheries, Generally,

August 1939 to 26 September 1939, LAC, RG 23, vol. 536, folder 711-25-24 (35).

6 "Be Patriotic! Serve Lobster: It's a Pleasant Way to Serve Canada," *Gazette* (Montreal), 22 July 1940, 5.

7 "Eat More Fish," *Guelph Daily Mercury*, 21 July 1941.

8 Ian Mosby, *Food Will Win the War: The Politics, Culture, and Science of Food on Canada's Home Front* (Vancouver: University of British Columbia Press, 2014), 3–5, 6, 8, 14.

9 Ibid., 25–38.

10 Ibid., 38–9, 42–51. Mosby does not specifically refer to Protestantism, but some of the works he cites do, and it seems clear that any assessment of pre-war Progressive reform would allow for the influence of the Protestant ideal and the power of individual choice in North American culture.

11 Ibid., 62.

12 Ibid., 61–96.

13 D.B. Finn, Deputy Minister, to W. George Akins, Walsh Advertising Company Ltd, Toronto, 20 December 1943, Advertising Fisheries, Generally, 29 August 1942 to 19 August 1947, LAC, RG 23, vol. 538, folder 711-25-24 (49).

14 W. George Akins, Vice-President, Walsh Advertising Company Ltd, to Hon. J.E. Michaud, Minister of Fisheries, 21 May 1942, Advertising Fisheries, Generally, 16 September 1941 to 25 June 1942, LAC, RG 23, vol. 538, folder 711-25-24 (48). After an initial rejection, Akins resubmitted essentially the same proposal on 15 August 1942; see ibid.

15 I.M. Dester's 1978 article suggests that there is nothing inherently bad about either humanitarian or realist approaches to food aid. Instead, he argues, "Food policy often appears contradictory and incoherent because it reflects a variety of conflicting, but entirely legitimate, values held by participants who have different stakes in the government's response to food and agriculture problems. Policy making inevitably involves trade-offs among these values. *Food policy*, therefore, is not a relatively autonomous policy area, manageable largely on its own terms, but an area where many interests and values converge and often compete." I.M. Dester, "United States Food Policy, 1972–1976: Reconciling Domestic and International Objectives," *International Organizations* 32, no. 3 (1978): 617–53. See also S.G. Triantis, "Canada's Interest in Foreign Aid," *World Politics* 24, no. 1 (1971): 1–18; Raymond F. Hopkins and Donald J. Puchala, "Perspectives on the International Relations of Food," *International Organizations* 32, no. 3 (1978): 581–616; Henry R.

Nau, "The Diplomacy of World Food: Goals, Capabilities, Issues and Arenas," *International Organization* 32, no. 3 (1978): 775–809; A.E. Laiou-Thomadakis, "The Politics of Hunger: Economic Aid to Greece, 1943–45," *Journal of the Hellenic Diaspora* 7, no. 2 (1980): 27–42; Theodore Cohn, *The Politics of Food Aid: A Comparison of American and Canadian Policies* (Montreal and Kingston: McGill-Queen's University Press, 1985); Mark W. Charlton, *The Making of Canadian Food Aid Policy* (Montreal and Kingston: McGill-Queen's University Press, 1992); Narayan Khadka, "U.S. Aid to Nepal in the Cold War Period: Lessons for the Future," *Pacific Affairs* 73, no. 1 (2000): 77–95; Ryan Macdonald and John Hoddinott, "Determinants of Canadian Bilateral Aid Allocations: Humanitarian, Commercial or Political?," *Canadian Journal of Economics/Revue Canadienne d'Economique* 37, no. 2 (2004): 294–312; Violetta Hionidou, "Relief and Politics in Occupied Greece, 1941–4," *Journal of Contemporary History* 48, no. 4 (2013): 761–83.

16 See, for example, Cranford Pratt, "Ethical Values and Canadian Foreign Aid Policies," *Canadian Journal of African Studies/Revue Canadienne des Études* 37, no. 1 (2003): 84–101.

17 Peter Urvin, "Regime, Surplus, and Self-Interest: The International Politics of Food Aid," *International Studies Quarterly* 36, no. 3 (September 1992): 293–312.

18 Harriet Friedmann, "The Political Economy of Food: The Rise and Fall of the Post War International Food Order," *American Journal of Sociology*, no. 88 (1982): S248–S286.

19 Ibid.

20 J.R. Tarrant, "The Geography of Food Aid," *Transactions of the Institute of British Geographers* 5, no. 2 (1980): 125–40.

21 Carmel Finley, *All the Fish in the Sea: Maximum Sustainable Yield and the Failure of Fisheries Management* (Chicago: University of Chicago Press, 2011), 2.

22 Ibid., 67–8.

23 Ibid., 51. American efforts to conserve fish extended only to territorial waters out to the three-mile exclusive economic zone (EEZ), which also encompassed large bays such as Chesapeake Bay in Maryland and Virginia or Bristol Bay in Alaska, where the headlands constituted the terrestrial basis for estimating the three-mile EEZ.

24 Donovan Finn was born in London, England, but received his education in Canada, with a BS from the University of Manitoba in 1924 and an MS in 1928. He received his PhD from Cambridge in 1932. Finn was in Prince Rupert from 1926 to 1935 before serving as the director of the

Halifax station until 1939 and then was chairman of the Salt Fish Board. He became the deputy minister of fisheries in 1940 and then the director of fisheries for the United Nations Food and Agriculture Organization in 1946. See Kenneth Johnstone, *The Aquatic Explorers: A History of Fisheries Research Board of Canada* (Toronto: University of Toronto Press, 1977).

25 Finley, *All the Fish in the Sea*, 146, 152.

26 Finley, *All the Boats on the Ocean*, 14, 60, 73, 96, 109, 119, 142.

27 Eric Roll, *The Combined Food Board: A Study in Wartime International Planning* (Stanford: Stanford University Press, 1956).

28 Newfoundland also expressed deep concern for the global shift in fisheries supply-and-demand relationships. See Governor of Newfoundland to P.J. Noel-Baler, P.C., M.P., Secretary of State for Commonwealth Relations, 6 December 1947, European Recovery Program, Canadian E.R.P. Supply Committee, 25 March 1948 to 11 August 1948, LAC, RG 23, vol. 1630, folder 792-5-7 (1).

29 D.B. Finn, Department of Fisheries, "Fish Products: Canadian Supplies and Requirements, 1942, Position," Produced on Behalf of the Food Requirements Committee, for the Combined Food Board, 1 December 1942, National Defense, Food Requirements Committee, 20 August 1942 to 11 June 1943, LAC, RG 23, vol. 1659, folder 792-24-1 (1).

30 "Draft: Interdepartmental Food Requirements Committee," 21 August 1942, National Defense, Food Requirements Committee, 20 August 1942 to 11 June 1943, LAC, RG 23, vol. 1659, folder 792-24-1 (1).

31 "Draft: Food Requirements Committee, 15 September 1942, Report of Prime Minister," National Defense, Food Requirements Committee, 20 August 1942 to 11 June 1943, LAC, RG 23, vol. 1659, folder 792-24-1 (1).

32 Minutes of Meeting Held in the Prime Minister's Office, 10:30 AM on September 16th, to Discuss the Relationship of Canada to Certain of the Combined Boards in Washington, National Defense, Food Requirements Committee, 20 August 1942 to 11 June 1943, LAC, RG 23, vol. 1659, folder 792-24-1 (1).

33 Ibid.

34 Ibid.

35 John Herd Thompson and Stephen Randall, *Canada and the United States: Ambivalent Allies*, 4th ed. (Athens: University of Georgia Press, 1994, 2008), 155.

36 Memorandum: Canadian Membership in the Combined Food Board, 23 July 1943, Food Requirements Committee, Generally, 26 July 1943 to 23 September 1943, LAC, RG 23, vol. 1660, folder 792-24-1 (6).

37　Press release for 28 October 1943, Food Requirements Committee, Generally, 28 October 1943 to 23 November 1943, LAC, RG 23, vol. 1660, folder 792-24-1 (8).

38　Combined Food Board Press Release, 30 July 1943, Food Requirements Committee, Generally, 26 July 1943 to 23 September 1943, LAC, RG 23, vol. 1660, folder 792-24-1 (6).

39　In all, Trinidad requested 2,464,000 pounds of butter for the coming 1944 year. See Meeting of the Sub-Committee on Dairy Products, Ottawa, 9 August 1943, Food Requirements Committee, Generally, 26 July 1943 to 23 September 1943, LAC, RG 23, vol. 1660, folder 792-24-1 (6).

40　Food Requirements Committee, Meeting Minutes from 25 January 1943, National Defense, Food Requirements Committee, 20 August 1942 to 11 June 1948, LAC, RG 23, vol. 1659, folder 792-24-1 (1).

41　Ibid.

42　Food Requirements Committee, Minutes of the Second Meeting, 12 November 1948, National Defense, Food Requirements Committee, 20 August 1942 to 11 June 1948, LAC, RG 23, vol. 1659, folder 792-24-1 (1).

43　Stewart Bates, Special Assistant to Deputy Minister, Department of Fisheries, to Commander Duncan MacTavish, Office of the Privy Council, Ottawa, 28 April 1943, Aid to Allies at War, Mutual Aid, 20 May 1945 to 22 January 1945, LAC, RG 23, vol. 1638, folder 792-5-1 (1).

44　Bill 76, An Act for Granting to His Majesty Aid for the Purpose of Making Available Canadian War Supplies to the United Nations, House of Commons of Canada, Fourth Session, Nineteenth Parliament, 7 George VI, 1943, as passed by the House of Commons, 14 May 1943.

45　Memo for Director of Commercial Services, 8 April 1944, Aid to Allies at War, Mutual Aid, 20 May 1945 to 22 January 1945, LAC, RG 23, vol. 1638, folder 792-5-1 (1).

46　P.H. Gordon, Chairman of the Executive Committee, Canadian Red Cross, to D.B. Finn, Deputy Minister, Department of Fisheries, Ottawa, 29 July 1941, National Defense, Fish for the Canadian Red Cross, 10 July 1941 to 29 February 1944, LAC, RG 23, vol. 1628, folder 792-5-3 (1).

47　W.H. Virgin, Secretary, National Purchasing Committee, Canadian Red Cross, to Frank E. Payson, Salmon Canners' Operating Committee, 19 March 1942 and A.L. Barry, Chief Supervisor, Halifax, NS, Department of Fisheries, to D.B. Finn, Deputy Minister, Department of Fisheries, 5 November 1942, National Defense, Fish for the Canadian Red Cross, 10 July 1941 to 29 February 1944, LAC, RG 23, vol. 1628, folder 792-5-3 (1).

48　Memorandum for Deputy Minister, Ottawa, 30 September 1943,

National Defense, Fish for the Canadian Red Cross, 10 July 1941 to 29
February 1944, LAC, RG 23, vol. 1628, folder 792-5-3 (1). See also Memorandum for the Deputy Minister, Ottawa, 1 May 1944, Canadian Red
Cross, Supply of Canned Fish for Canadian Red Cross, 25 March 1944
to 11 August 1946, LAC, RG 23, vol. 1628, folder 792-5-3 (2).

49 Statement Showing Quantity of Sardines Shipped by Connors Bros.,
Ltd, Black Harbour, N.B., January–September 1944, Canadian Red
Cross, Supply of Canned Fish for Canadian Red Cross, 25 March 1944
to 11 August 1946, LAC, RG 23, vol. 1628, folder 792-5-3 (2).

50 Department of Fisheries, Principles Underlying Issuance of Export Permits, Fish Commodities, Memorandum for the Food Requirements
Committee, Second Meeting, Minute no. 2, November 12th, 25 November 1942, National Defense, Food Requirements Committee, 20 August
1942 to 11 June 1948, LAC, RG 23, vol. 1659, folder 792-24-1 (1).

51 In 1943, the Department of Fisheries' planned distribution included 319
million pounds for domestic consumption, 320 million pounds for
export to the United States, 191 million pounds for export to the United
Kingdom, and 110 million pounds for export to all other nations, leaving an estimated 100 million pounds of stock on hand as of 1 January
1944. Review of Canada's Food Position, Fisheries, Food Requirements
Committee, Food Requirements Committee, Generally, 30 November
1943 to 17 January 1944, LAC, RG 23, vol. 1660, folder 792-24-1 (9).

52 Canada's Role in UNRRA, Wartime Information Board, Ottawa, no. 18,
23 December 1943, Supply of Fish for Relief, through UNRRA, 22 April
1944 to 13 September 1944, LAC, RG 23, vol. 1629, folder 792-5-5 (1).

53 Memorandum for Mr Angus, Re: Canned Fish for UNRRA, Ottawa, 3
June 1944, Supply of Fish for Relief, through UNRRA, 22 April 1944 to
13 September 1944, LAC, RG 23, vol. 1629, folder 792-5-5 (1).

54 G.L. Peterson, United Nations Relief and Rehabilitation Administration, Washington, to A. Cairns, Food Division, United Nations Relief
and Rehabilitation Administration, 12 August 1944, and G.L. Peterson,
Food Division, to Miss Jane Harding, Balkan Division, Liberated Areas
Branch, Foreign Economic Administration, Washington, DC, 11 August
1944, Supply of Fish for Relief, through UNRRA, 22 April 1944 to 13
September 1944, LAC, RG 23, vol. 1629, folder 792-5-5 (1).

55 A. Cairns, Chief, Food Division, to Major Hugh A Green, Vice President
and Managing Director, Norwegian-Canadian Packers Ltd, Norway,
New Brunswick, 11 August 1944, Supply of Fish for Relief, through
UNRRA, 22 April 1944 to 13 September 1944, LAC, RG 23, vol. 1629, folder 792-5-5 (1).

56 Clive Planta, Secretary-Manager, Fisheries Council of Canada, Ottawa, to
Ian MacArthur, Chief Economist, Department of Fisheries, Ottawa, 21
January 1948, Supply of Fish for United Nations Relief, 13 January 1948
to 16 January 1952, LAC, RG 23, vol. 1630, folder 792-5-5 (10).

57 S.H. Burhoe, Manager, J.W. Windsor Co. Limited, Charlottetown, PEI, to
Stewart Bates, Acting Chairman Fisheries Price Support Board, Ottawa,
5 March 1948, Supply of Fish for United Nations Relief, 13 January
1948 to 16 January 1952, LAC, RG 23, vol. 1630, folder 792-5-5 (10).

58 Exchange of letters included in Supply of Fish for United Nations
Relief, 29 August 1947 to 7 January 1948, LAC, RG 23, vol. 1630, folder
792-5-5 (9).

59 Committee on Agriculture, Europe, United Nations Relief and Rehabilitation Administration, Fisheries Rehabilitation and Relief Supplies of Fish, 8 October 1944, Supply of Fish for Relief, UNRRA, 14
September 1944 to 14 November 1944, LAC, RG 23, vol. 1629, folder
792-5-5 (2).

60 The UNRRA spelled out the machinery and equipment needed to
restore European fishing capacity, in Subcommittee on Agriculture for
Europe, Expert Panel on Fisheries, Memorandum on Fish Production
and Consumption and the Rehabilitation of Fisheries, Supply of Fish
for United Nations Relief, UNRRA, 16 November 1944 to 16 February
1945, LAC, RG 23, vol. 1629, folder 792-5-5 (3).

61 UNRRA Requirements from all Sources for First Six Months after the
Military Period, included in A. Cairns, Chief, Food Division, to George
R. Paterson, Commercial Attache, Canadian Embassy Annex, Washing-
ton, DC, 22 November1944, Supply of Fish for United Nations Relief,
UNRRA, 16 November 1944 to 16 February 1945, LAC, RG 23, vol. 1629,
folder 792-5-5 (3).

62 Teletype from the Canadian Ambassador to the United States to the
Secretary of State for External Affairs, Canada, Washington, 11 June
1946, Supply of Fish for United Nations Relief, UNRRA, 9 March 1946 to
20 August 1946, LAC, RG 23, vol. 1629, folder 792-5-5 (6).

63 H.A. Gilbert, Export Division, Foreign Trade Service, Ottawa, to E.J.
Wadley, Canadian Commercial Corporation, Ottawa, 18 July 1946, and
Andrew Cairns, Director, Food Division, United Nations Relief and
Rehabilitation Administration, Washington, to L.B. Pearson, Canadian
Embassy, Washington, 24 July 1946, Supply of Fish for United Nations
Relief, UNRRA, 9 March 1946 to 20 August 1946, LAC, RG 23, vol. 1629,
folder 792-5-5 (6).

64 The arrangement is outlined in numerous letters in Supply of Fish for

United Nations Relief, UNRRA, 20 August 1946 to 24 June 1947, LAC, RG 23, vol. 1630, folder 792-5-5 (7).

65 Clive Planta, Secretary-Manager, Fisheries Council of Canada, Ottawa, Memorandum no. 47-47, to Secretaries and Members of Member Organizations, 24 June 1947, Supply of Fish for United Nations Relief, UNRRA, 24 June 1947 to 28 August 1947, LAC, RG 23, vol. 1630, folder 792-5-5 (8).

66 Combined Food Board, Second Report to United States, Canadian, and United Kingdom Members of the Council of the United Nations Relief and Rehabilitation Administration, London, England, 7 August 1945, Supply of Fish for United Nations Relief, UNRRA, 8 August 1945 to 27 February 1946, LAC, RG 23, vol. 1629, folder 792-5-5 (5).

67 Teletype, Canadian Ambassador to the United States to the Secretary of State for External Affairs, Canada, Washington, 13 June 1947, signed, Charge D'Affaires, Supply of Fish for United Nations Relief, UNRRA, 20 August 1946 to 24 June 1947, LAC, RG 23, vol. 1630, folder 792-5-5 (7).

68 Ian McArthur, Chief Economist, Department of Fisheries, Memorandum for Deputy Minister, Re Fish in European Program, 29 January 1948, European Recovery Program, Canadian E.R.P. Supply Committee, 25 March 1948 to 11 August 1948, LAC, RG 23, vol. 1630, folder 792-5-7 (1). See also Ian McArthur, Chief Economist, Department of Fisheries, to S.M. Rosenberg, Chairman, Salmon Canners' Operating Committee, Vancouver, 30 April 1948, European Recovery Program, Canadian E.R.P. Supply Committee, 25 March 1948 to 11 August 1948, LAC, RG 23, vol. 1630, folder 792-5-7 (1).

69 Governor of Newfoundland, to P.J. Noel-Baler, P.C., M.P., Secretary of State for Commonwealth Relations, 6 December 1947, European Recovery Program, Canadian E.R.P. Supply Committee, 25 March 1948 to 11 August 1948, LAC, RG 23, vol. 1630, folder 792-5-7 (1).

70 Draft Memorandum of the United Nations for Freedom from Want of Food, October 1942, sent by L.B. Pearson, 19 November 1942, to Secretary of State for External Affairs, Ottawa, National Defense, Food Requirements Committee, 20 August 1942 to 11 June 1948, LAC, RG 23, vol. 1659, folder 792-24-1 (1).

71 Coveney, Food Morals and Meaning; Ziegelman, A Square Meal; Callather, "The Foreign Policy of Calories"; James, Hunger.

72 Draft Memorandum of the United Nations for Freedom from Want of Food, October 1942, sent by L.B. Pearson, 19 November 1942, to Secretary of State for External Affairs, Ottawa, National Defense, Food Requirements Committee, 20 August 1942 to 11 June 1948, LAC, RG 23, vol. 1659, folder 792-24-1 (1).

73 Teletype, Canadian Minister in the United States to the Secretary of State for External Affairs, Washington, 10 January 1944, United Nations Interim Commission on Food and Agriculture, January 1944 to 5 June 1944, LAC, RG 12, vol. 1673, folder, 792-24-(1).

74 United Nations Interim Commission on Food and Agriculture, Summary of Special Meeting of the Reviewing Panel with Chairmen and Rapporteurs of the Panel Subcommittees Held on 1 August 1944, Washington, DC, United Nations Interim Commission on Food and Agriculture, August 1944 to 9 September 1944, LAC, RG 23, vol. 1674, folder 792-24-7 (4).

75 United Nations Interim Commission on Food and Agriculture, Outline for General Report of the Reviewing Panel, First Draft, 30 November 1944, United Nations Interim Commission, Food and Agriculture, Generally, 1 December 1944 to 8 January 1945, LAC, RG 23, vol. 1675, folder 792-24-7 (10).

76 United Nations Interim Commission on Food and Agriculture, General Report of the Reviewing Panel, First Draft, 12 January 1945, 29, United Nations Interim Commission, Food and Agriculture, Generally, 1 December 1944 to 8 January 1945, LAC, RG 23, vol. 1675, folder 792-24-7 (10).

77 Ibid., 30.

78 United Nations Interim Commission, Food and Agriculture, Subcommittee on Agricultural Products, doc. 5, draft, Report of the Subcommittee on Agricultural Products, 1944, United Nations Interim Commission, Food and Agriculture. Generally, 11 September 1944 to 19 September 1944, RG 23, vol. 1674, folder 792-24-7 (7).

79 Ian McArthur, Chief Economist, Department of Fisheries, Memorandum for Deputy Minister, Re Government Buying of Fish, 29 December 1947, European Recovery Program, Canadian E.R.P. Supply Committee, 25 March 1948 to 11 August 1948, LAC, RG 23, vol. 1630, folder 792-5-7 (1).

80 Clive Planta, Secretary-Manager, Fisheries Council of Canada, to Mr Stewart Bates, Deputy Minister of Fisheries, Ottawa, 6 April 1948, Advertising Fisheries, Generally, 20 August 1947 to 23 February 1951, LAC, RG 23, vol. 539, folder 711-25-25 (50).

81 Clive Planta, Secretary-Manager, Fisheries Council of Canada, to Stewart Bates, Deputy Minister of Fisheries, 18 October 1948, Advertising Fisheries, Generally, 20 August 1947 to 23 February 1951, LAC, RG 23, vol. 539, folder 711-25-25 (50).

82 Committee of Advertising, Report of a Meeting Held Wednesday Afternoon, 13 November 1947, Expansion of Fishing Trade, Committee on Advertising, 13 November 1948 to March 1953, LAC, RG 23, vol. 541, folder 711-25-35 (1).

83 O.C. Young, Memorandum on Fisheries Development in Canada, Ottawa, January 1949, Information, Expansion and Development of the Fisheries, 11 May 1948 to 9 August 1949, LAC, RG 23, vol. 541, folder 711-23-36 (1).

84 Mayhew was elected Liberal MP for Victoria in 1937 and served as minister of fisheries from 1948 to 1952 before becoming Canada's first ambassador to Japan.

85 Press release, announcement of Hon. R.W. Mayhew, Minister of Fisheries, Ottawa, 5 May 1949, Information, Expansion and Development of the Fisheries, 11 May 1948 to 9 August 1949, LAC, RG 23, vol. 541, folder 711-23-36 (1). More and more of the documents related to the expansion of the fisheries industry relate to inspection, storage, refrigeration, and other strategies to improve the quality of the product rather than advertising to consumers. See Information, Expansion and Development of the Fisheries, 10 August 1949 to October 1949, LAC, RG 23, vol. 541, folder 711-23-36 (2).

86 Memorandum Re Activities, 1940–1952 Department of Fisheries – Fisheries Research Board, 11 August 1952, Information, Expansion and Development of the Fisheries, October 1949 to 16 December 1952, LAC, RG 23, vol. 541, folder 711-23-36 (3).

87 G.R. Clark, Deputy Minister, Department of Fisheries, to E. Bossé, Executive Assistant to the Minister of Labour, 15 November 1955, Advertising Fisheries, Generally, 5 March 1951 to 17 July 1961, LAC, RG 23, vol. 539, folder 711-25-25 (51).

88 "Every Time a Canadian Housewife Reaches for Fish – She Puts Money into *Your* Pocket!," advertisement by the Department of Fisheries, *Western Business and Industry* 36, no. 12 (December 1962): 81, emphasis in original; also published in *Atlantic Advocate*, October 1962, 34.

89 The regional focus of the department's advertising is documented in numerous correspondences of T.H. Turner, director, Information and Consumer Services, Department of Fisheries, to newspapers seeking advertisements from the department. Turner regularly responded to them stating that the department only advertised in fishing regions where their message was most likely to be seen by those engaged in the industry. The clear focus of the department was reaching those who pro-

duced fish products rather than those who consumed them. See Advertising Fisheries, Generally, 24 September 1963 to 15 March 1964, LAC, RG 23, vol. 539, folder 711-25-25 (56).

90 Hon. J. Angus Maclean, MP, Minister, George R. Clark, Deputy Minister, "Big Business – Getting Bigger!," Department of Fisheries, Ottawa, Canada, advertisement, 1961, Advertising Fisheries, Generally, 5 March 1951 to 17 July 1961, LAC, RG 23, vol. 539, folder 711-25-25 (51).

91 "When Is the Best Time to Put Fish on *Your* Menu?," advertisement of Department of Fisheries, *Western Business and Industry*, July 1965.

92 H.J. Robichaud, Minister of Fisheries, to Bruno Tenhunen, President, Canada Ethnic Press Federation, Toronto, 26 May 1964, Advertising Fisheries, Generally, 16 March 1964 to 31 July 1964, LAC, RG 23, vol. 539, folder 711-25-25 (57). See also Memorandum Re Activities, 1940–1952 Department of Fisheries – Fisheries Research Board, 22 August 1952, Information, Expansion and Development of the Fisheries, October 1949 to 16 December 1952, LAC, RG 23, vol. 541, folder 711-23-36 (3).

93 Bates became deputy minister in 1942 after having taught commerce at Dalhousie University for four years. He also served as secretary of the Economic Council of Nova Scotia beginning in 1936. See his obituary in the *New York Times*, 25 May 1964, 33.

94 Report on the discussion held at Ottawa, 30 June and 1 July 1947, between representatives of the Dominion Department of Fisheries and the fisheries industry of the East Coast, Inland, and Pacific Coast, Supply of Fish for United Nations Relief, UNRRA, 24 June 1947 to 28 August 1947, LAC, RG 23, vol. 1630, folder 792-5-5 (8).

95 Letter from B. McInerney, Salt Fish Administration, Department of Fisheries, Ottawa, to I.S. McArthur, Chief Economist, Department of Fisheries, Ottawa, 21 June 1947, Supply of Fish for United Nations Relief, UNRRA, 24 June 24 1947 to 28 August 1947, LAC, RG 23, vol. 1630, folder 792-5-5 (8).

Index

athleticism and food advertising, 59, 210n50

Atlantic Canada, 8; economic depression, 8, 20, 44–6, 83, 111, 115, 191; government investigations on, 125, 137; lobster fisheries, 153; objection to National Policy, 24, 43–4; in relation to Maritime Rights Movement, 46–7; subsidies in support of, 113–14; trade with United States, 24, 43. *See also* trawlers; United Maritime Fishermen

Barry, A.L., 157, 241n149

Bell, Marjorie, 52–3, 163

Bennett, Richard Bedford, 134, 135–6

Black Brothers, 54, 55, 56–7, 63

brand-name marketing, 54, 55, 56–8, 81, 133, 156, 190

British Columbia Packers Association, 7, 53, 127–8. *See also* salmon, canned Pacific

Brittain, Alfred Herbert, 59, 100, 129, 140–1

Burnaby, R.W.E., 129, 232n62, 233n63

calories in food, 6, 15, 18, 92; highlighted in advertisements, 58

Campbell, Mildred Helena, 63–4, 69, 70–1, 72, 80, 103, 109, 130

Canada Food Board, 27, 28, 36, 38–9, 40, 41–2, 101, 109

Canadian Council on Nutrition (CNN), 52

Canadian Department of (Marine and) Fisheries, 6; on debate of healthfulness of seafood, 10, 49, 67, 73, 78; funding to private marketing, 8, 78, 109–11, 116, 118; on publication of seafood cookbooks, 85, 87, 92; on stimulating consumption, 27, 63, 69–70, 87. *See also* inspection of fish

Canadian Department of Pensions and Public Health, 52, 79

Canadian Fisheries Association, 4, 6, 7–8; calls for government funding, 113–14, 118–19, 126–7, 140–1,